The ChatGPT Millionaire Blueprint

Build A Successful Business And Create Financial Freedom With AI

Business Maven

SourceLife Publishing

Copyright © 2023 by Business Maven.

All rights reserved.

No portion of this book may be reproduced in any form without written permission from the publisher or author, except as permitted by U.S. copyright law.

This publication is designed to provide accurate and authoritative information in regard to the subject matter covered. It is sold with the understanding that neither the author nor the publisher is engaged in rendering legal, investment, accounting or other professional services. While the publisher and author have used their best efforts in preparing this book, they make no representations or warranties with respect to the accuracy or completeness of the contents of this book and specifically disclaim any implied warranties of merchantability or fitness for a particular purpose. No warranty may be created or extended by sales representatives or written sales materials. The advice and strategies contained herein may not be suitable for your situation. You should consult with a professional when appropriate. Any reliance you place on such information is there for strictly at your own risk. Neither the publisher nor the author shall be liable for any loss of profit or any other commercial damages, including but not limited to special, incidental, consequential, personal, or other damages.

THE CHATGPT MILLIONAIRE BLUEPRINT

SourceLife
Publishing

SourceLife Publishing, PO Box 585, Beverly Hills, CA 90213, USA

www.sourcelifepublishing.com

Contents

Dedication	VII
Title Page	IX
Blurbs	X
1. WELCOME	1
3. HOW TO USE THIS BOOK	10
4. INTRODUCTION TO CHATGPT	12
6. MARKET RESEARCH AND NICHE SELECTION	21
8. BUSINESS MODEL AND REVENUE STREAMS	50
10. LAYING THE GROUNDWORK FOR YOUR CHATGPT BUSINESS	74
12. CHATGPT INTEGRATION AND OPTIMIZATION	107
14. CONTENT CREATION AND AUTOMATION	129

16.	MARKETING AND PROMOTION	153
19.	CUSTOMER SUPPORT AND RETENTION	181
21.	MONITORING AND ANALYTICS	202
23.	SCALING YOUR CHATGPT BUSINESS	221
24.	CONCLUSION	237
BONUS - PROMPTS		241
ABOUT AUTHOR		244
With Thanks		246

To my partner, my business partner, my best friend, the original Mr. Blueprint, Richard Krawczyk who passed away suddenly in November 2022. This book and its title are my homage to you.

You were my world, and your passing forced me to reassess my purpose in life. Losing you was the catalyst for me rediscovering my passion for consulting and coaching businesses and fueling my exploration of the power of Artificial Intelligence to help mankind live to its full potential. Although my heart is heavy as I write this, I know you are watching and cheering me on as only you could.

Thank you for showing me what I am capable of.

With all my love.

The ChatGPT Millionaire Blueprint

Build A Successful Business And Create Financial Freedom With AI

Business Maven

SourceLife Publishing

Blurbs

"The ChatGPT Millionaire Blueprint" is the indispensable guide for ambitious people who want to capitalize on the game-changing potential of AI, which is poised to create more millionaires in the next two years than in the entire last century.

In this captivating and practical step-by-step manual, veteran Fortune 500 Consultant and AI Strategist, Business Maven; offers a series of insightful chapters, empowering you to:

- Grasp the transformative impact of AI and ChatGPT on business, and how it's removing barriers and obstacles that traditionally led to small business failures.

- Establish a profitable ChatGPT-based business you're passionate about. And make it happen NOW.

- Generate impressive revenue streams you've never imagined, utilizing the power of AI.

By the end of "The ChatGPT Millionaire Blueprint," you'll understand the unprecedented opportunities presented by ChatGPT, how to embrace and capitalize on AI's potential, and how to master the power of ChatGPT to build a business with unparalleled success and join the new wave of AI millionaires!

WELCOME

A LITTLE BACKGROUND BEFORE WE START

In nearly 30 years of working in Business. I have witnessed all sorts of booms and busts in the world economy.

I remember a world of analog where professionals could only get their work done with the help of secretaries and manual typewriters, the launch of the Internet, and the.com bubble bursting. Crypto being the hottest thing to land in the world of digital and of course the financial crisis in the banking system in 2008.

There have been so many economic lessons over the last 30-odd years for me. A career that has seen me work with over 1,000 small business owners and many of the largest organizations in the world.

I have helped many global organizations move their operations from analog to digital, specializing in Digital Transformation. I also had the intriguing experience of implementing artificial intelligence into one of the largest banks in the world and witnessing the power of AI in the corporate world pre-pandemic.

With all of this experience, I can tell you hand on heart when I saw what artificial intelligence is capable of doing today with the launch of Open AI's ChatGPT, I was blown away and I'm not the only one.

When ChatGPT went live in November 2022 it gained 1 million users in just 5 days. In context, it took Netflix 3 years, Twitter 2 years, and Facebook 10 months to hit 1 million users. As of March 2023, the number of users of OpenAI's ChatGPT exceeded 100 million users. And the world has not yet fully woken up to its potential.

For the first time in my entire professional life, I believe we are witnessing a revolution in the world of not just work but in human existence.

And as a student of history, exploring the capabilities of ChatGPT, I could not help but draw comparisons to the invention of the printing press and the revolution that this simple piece of equipment caused and the subsequent impact on learning and religion worldwide.

But don't just take my word for it. In a report from Goldman Sachs dated March 2023, they estimated over the next few years over 300 million jobs will be impacted by AI, which is a very scary thought. And despite the likes of Elon Musk & Steve Wozniak signing a letter calling for the halt to any AI development beyond GPT 4 (which is where we are currently in May 2023) AI is not going anywhere fast.

It is here to stay, and it is growing and evolving every day. I often joke that if you miss one day in AI, it feels like you've missed a month in learning. The rate of growth of AI is exponential, and it is mind-blowing.

But despite the fear-mongering, I believe this is one of the biggest opportunities we will witness in our lifetimes.

Open AI and this wave of artificial intelligence that is flooding the market right now is bigger than any other opportunity that has come into our lives this century. Mark my words **AI will create more millionaires in the next 2 years than we have seen in the entire last century!**

The opportunity

As a Fortune 500 consultant and Small Business Coach, I am well aware of the barriers to launching and running a successful business. The statistics around survival rates of small businesses are so shocking it's a wonder anyone ever tries to launch a business.

But as I explored the knowledge of ChatGPT, I was astounded at how robust the ideas and solutions that came from this Large Language Model (LLM). I believe ChatGPT and the other LLMs out there have **dramatically cut down the risks** of building and launching a successful business. Need a Marketing expert? Ask AI. Need a Sales Pitch? Ask AI, need a plan for the week ahead. Ask AI! AI can take the place of almost any member of a team – although currently it is not connected to the internet. But it will be by the end of 2023!

So with a combination of my experience as a Fortune 500 Consultant and Small Business Coach, combined with my knowledge of implementing AI and working with ChatGPT I decided to create this step-by-step guide to help anybody get through the barriers of creating a successful business by using AI.

The opportunity to create something right now at the start of this wave is phenomenal and it is exciting I believe that **anybody with enough will and enough passion has the opportunity at their fingertips right now to create the life of their dreams by**

simply following the guidance in this book.

Through a combination of templates, top tips, and prompts all you need is an account with any of the AI providers, the will, and this book to create the life you deserve to live.

In **The ChatGPT Millionaire Blueprint**, I am dedicated to sharing my expertise and passion for AI, and helping entrepreneurs like you, succeed in today's competitive marketplace. With the knowledge and tools provided in this comprehensive guide, you'll be equipped to build a ChatGPT-based business that capitalizes on AI's immense potential and drives measurable growth.

Throughout this guide, you will:

1. Identify your target audience and niche, and assess market demand and competition

2. Develop your ChatGPT-based business model and discover potential revenue streams

3. Integrate ChatGPT seamlessly into your business, optimize its performance, and minimize costs

4. Create and automate high-quality, engaging content using ChatGPT's powerful capabilities

5. Market and promote your business effectively, leveraging ChatGPT to enhance your efforts

6. Improve customer support and retention with the help of ChatGPT's natural language understanding

7. Monitor and analyze your business's performance, making data-driven decisions for success

8. Scale your ChatGPT business, identifying growth opportunities and expanding efficiently

If you're ready to harness the power of AI and expert guidance to build a successful ChatGPT-based business, dive into The ChatGPT Millionaire Blueprint and discover a wealth of actionable insights and practical strategies to help you achieve your entrepreneurial goals.

I encourage you to dive in, explore the wealth of resources provided, and take the first steps toward transforming your business with ChatGPT. Remember, the only limit to your success is your imagination.

Everybody has the right to live life. You deserve to live your life to the fullest, and I believe that the power of AI will empower every single one of us to be able to achieve that dream! So without further ado, enjoy this book and I hope that you will share your story with me in the future.

> "Happiness depends upon ourselves."

Aristotle, Philospher

HOW TO USE THIS BOOK

Each section is broken down into categories for building a business from scratch. There is no one size fits all as it's entirely dependent on your experience, desires, capabilities, and appetite. So I have put together the foundations of each section. You'll find an introduction, and a section of templates for you to work through in your own time, to ensure you answer all the essential questions and fully develop your business plan.

There are also some top tips and checklists in some of the chapters to get your creative juices flowing as you work through this book.

Finally, at the end of each section I have included a copy-and-paste ChatGPT prompt as well as an example of what that prompt would look like for an example business.

Now I know some of you will just jump straight into the ChatGPT copy-and-paste prompts but please bear in mind that without the additional information in each chapter, there may be some areas of your plan missing.

I would advise that you work through each section fully and use the prompts with ChatGPT to help you create the most solid business foundation you can.

But it's ultimately up to you how you wish to proceed. If you do need any further assistance or support in your business please get in touch with me at my website www.businessmaven.io.

Here's to your success!

INTRODUCTION TO CHATGPT

REVOLUTIONIZING AI FOR BUSINESS

Since its launch in late 2022, ChatGPT has taken the world by storm, revolutionizing how people interact with artificial intelligence. This groundbreaking AI LLM offers a multitude of functions, including holding conversations, generating content, designing logos, and even composing music. For entrepreneurs and business owners, understanding ChatGPT and its capabilities can be a game-changer.

This chapter will provide an in-depth introduction to ChatGPT, its capabilities, and how you can harness its power for your business. We'll address frequently asked questions and important considerations for first-time users. Get ready to explore the world of ChatGPT and unlock its potential for your business.

What is ChatGPT?

ChatGPT is an AI language model designed to generate human-like text and engage in conversations. To interact with ChatGPT, you simply input a query or request into a text box, and the AI processes the information and responds accordingly.

Key capabilities of ChatGPT include:

1. Generating written content, from news articles to novels

2. Summarizing lengthy documents

3. Acting as a research assistant by answering questions

4. Writing and debugging code

5. Developing text-based games

6. Serving as a tutor for homework questions or problems

7. Assisting with travel planning

8. Creating software activation keys

Understanding GPT Models

The "GPT" in ChatGPT refers to the Generative Pre-trained Transformer model used by the chatbot. Currently, most users interact with GPT-3.5, which powers the free research preview version of ChatGPT. However, a more advanced GPT-4 model is available exclusively to ChatGPT Plus subscribers and developers utilizing the GPT-4 API. (As of May 2023)

Free and Paid Access to ChatGPT

ChatGPT remains available as a free service during its research phase. To access it, simply create an account at www.openai.com and decide on whether you want the free or paid-for plan. **The ChatGPT Millionaire Blueprint** works with both plans.

OpenAI also offers ChatGPT Plus, a $20/month subscription service that includes faster response times, priority access to new features, and the more advanced GPT-4 model. However, be aware that some users have reported difficulties accessing ChatGPT due to high demand.

Developers can leverage the ChatGPT API to integrate AI into their apps, which operate on a pay-as-you-go pricing model. Additionally, opportunities for full-time employment as a Prompt Engineer or participating in a bug bounty program to find and report ChatGPT issues are available.

ChatGPT Restrictions and Availability

As of now, ChatGPT is banned in Italy due to

concerns over data collection and age verification. However, this ban may be temporary if OpenAI addresses Italy's concerns. ChatGPT is also unavailable in China, Russia, North Korea, and Iran.

When using ChatGPT, it's important to be vigilant about plagiarism, as the AI may unintentionally use text from online sources without proper citation. You have been warned!

Comparing GPT Models: GPT-3.5 and GPT-4

The primary differences between GPT-3.5 (ChatGPT) and GPT-4 (ChatGPT-4) lie in their processing capabilities and available features. GPT-4 is a significant upgrade, offering faster processing, the ability to handle more lines of text and even image processing. GPT-4's multimodal functionality allows it to work with both text and images, while GPT-3.5 is limited to text-based inputs and outputs.

Future GPT Models: ChatGPT-5

While ChatGPT-4 is already an impressive AI language model, researchers at OpenAI are continuously working on the next generation of language models. It is anticipated that ChatGPT-5, the successor to ChatGPT-4, will bring even more advanced capabilities to the table. Although specific details about its release and improvements are not available at this time, we can still expect some general enhancements in ChatGPT-5:

1.Improved language understanding:

- As AI language models continue to improve, we can expect ChatGPT-5 to have an even better understanding of natural language, allowing it to provide more accurate and contextually relevant responses.

2. Multilingual support:

- ChatGPT-5 may expand its language capabilities, supporting more languages and improving its performance for non-English speakers.

3. Multimodal integration:

- Future GPT models may be able to integrate with other AI technologies, such as computer vision and speech recognition, to provide a more comprehensive and interactive user experience.

4. Enhanced safety and ethical considerations:

- OpenAI says that they are committed to addressing the ethical challenges posed by AI language models. ChatGPT-5 may incorporate new safety features and improvements to address issues like harmful outputs and biases in the model.

5. Customizability:

- ChatGPT-5 could potentially offer more customization options for businesses, allowing them to tailor the AI's behavior and responses to better align with their specific needs and values.

It's important to note that these are speculative improvements, and the actual features of ChatGPT-5 may differ from what's mentioned above. As AI continues to advance at a rapid pace, businesses must stay up-to-date with the latest developments to maximize the benefits of these powerful tools.

But now you have the essentials, let's dive into building the business of your dreams.

"The size of your success is measured by the strength of your desire; the size of your dream; and how you handle disappointment along the way."

Robert Kiyosaki, American Entrepreneur & Author

MARKET RESEARCH AND NICHE SELECTION

UNEARTHING PROFITABLE OPPORTUNITIES

T he foundation of any successful business lies in its ability to effectively identify and cater to the needs of its target audience. For ChatGPT-based businesses, this principle remains unchanged. By honing in on a well-defined niche and understanding the preferences of your target audience, you'll be better positioned to develop offerings and messaging that resonate with them. In this section, we'll explore the importance of market research and niche selection, and how these crucial steps can set the stage for your ChatGPT-based business to thrive.

In today's rapidly evolving business landscape, consumers are bombarded with countless products, services, and marketing messages. To cut through the noise and grab their attention, your business needs to offer something that is both unique and relevant to their needs. This is where the importance of selecting the right target audience and niche comes into play. By focusing on a specific group of people and addressing a well-defined need or problem, you can create a competitive edge for your ChatGPT-based business.

Market Research

Market research is a valuable tool that can help you gain insights into your target audience and niche. By analyzing data on consumer behavior, preferences, and industry trends, you can better understand the dynamics of your chosen market. This knowledge will empower you to make informed decisions about your business strategy, from the development of your ChatGPT-based solutions to the design of your marketing campaigns.

Another critical aspect of market research is the identification of your competitors. By examining the strengths and weaknesses of other businesses operating within your niche, you can uncover opportunities for differentiation and innovation. This will enable you to create a unique selling proposition (USP) that sets your ChatGPT-based business apart from the competition, making it more attractive to your target audience.

The process of selecting the right target audience and niche is not a one-time event. As consumer needs and preferences evolve, and as new competitors enter the market, it's essential to continually reassess your position within the niche. This ongoing analysis will ensure that your ChatGPT-based business remains relevant and competitive in the long run.

In summary, the importance of selecting the right target audience and niche for your ChatGPT-based business cannot be overstated. By conducting thorough market research and continually refining your understanding of your target audience's needs and preferences, you can create a strong foundation for your business. This will ultimately enable you to develop innovative and relevant solutions that not only attract attention but also foster long-term customer loyalty.

The Niche

By focusing on a well-defined niche and understanding your target audience's needs and preferences, you can tailor your offerings and messaging for maximum impact. This section will provide you with tools and insights to help you make informed decisions about your market and niche.

WORKSHEET:

IDENTIFYING YOUR TARGET AUDIENCE & NICHE

To begin, complete the following worksheet to clarify your target audience and niche:

1. Define your Audience and Niche:

- Define your target audience:

- Demographics (age, gender, location, etc.)

- Interests and hobbies

- Pain points and challenges

- Preferred communication channels

2. Identify your niche:

- What specific problem or need does your ChatGPT-based business address?

- How does your solution differ from existing alternatives?

- What value do you provide that sets you apart from competitors?

3. Evaluate your passion and expertise in the niche:

- How familiar are you with the niche and its audience?

- Do you have any unique insights or experiences that can contribute to your success in this niche?

Overall, by completing this worksheet, you have a clearer understanding of your target audience and niche, as well as your unique value proposition. This can help guide your business strategy, marketing efforts, and overall success in the marketplace. Let's take a look at the example below for inspiration.

EXAMPLE:

Let's say you're starting a company that uses AI to personalize meal plans for people with specific dietary needs. Here's how you could fill out the worksheet:

1.Define your target audience:

- **Demographics:** health-conscious individuals, primarily aged 25-45, living in urban areas in the United States.

- **Interests and hobbies:** fitness, healthy cooking, and nutrition.

- **Pain points and challenges:** struggling to find meal plans that meet their dietary restrictions, limited time to plan and prepare meals, and desire for variety in their diet.

- **Preferred communication channels:** social media (e.g., Instagram, Facebook), email, and in-app messaging.

2. Identify your niche:

- **Problem or need:** There is a lack of personalized meal plans that meet the dietary restrictions of health-conscious individuals with busy lifestyles.

- **Solution differentiation:** Our AI-based meal planning algorithm uses individualized data (e.g., body metrics, dietary restrictions, and preferences) to create personalized meal plans that offer variety, flexibility, and convenience. Unlike traditional meal planning services, we use AI to constantly adjust and optimize meal plans to meet our client's changing needs.

- **Value proposition:** Our personalized meal plans provide our clients with a convenient, time-saving solution that addresses their unique dietary needs and preferences, leading to improved health outcomes and overall satisfaction.

3. Evaluate your passion and expertise in the niche:

- **Familiarity:** Our team has a deep understanding of nutrition and dietary restrictions, as well as experience in software development and data analysis.

- **Unique insights:** We have identified a gap in the market for personalized meal plans that meet the specific dietary needs of health-conscious individuals with busy lifestyles. Additionally, our AI-based approach allows us to provide ongoing optimization and customization, which sets us apart from traditional meal-planning services.

CHECKLIST:

Once you have a clear understanding of your target audience and niche, use the following checklist to assess market demand and competition:

1.Perform keyword research:

- Use tools like Google Keyword Planner, Ubersuggest, or Ahrefs to identify popular keywords in your niche. Use them to analyze search volume, competition, and cost per click (CPC) to gauge market interest

2. Analyze competitors:

- Identify your main competitors and evaluate their strengths and weaknesses Examine their product or service offerings, pricing, and marketing strategies

3. Assess market trends:

- Use resources like Google Trends, industry reports, and expert blogs to stay informed about market developments and emerging opportunities.

TOP TIPS:

USING CHATGPT TO DISCOVER MARKET INSIGHTS AND TRENDS

Leverage ChatGPT's powerful language understanding capabilities to gather valuable market insights and stay ahead of industry trends:

1. Generate content ideas:

- Use ChatGPT to brainstorm content topics related to your niche, helping you identify areas of interest to your target audience

2. Analyze customer feedback:

- Utilize ChatGPT to process and summarize customer reviews, comments, and feedback, revealing insights about your audience's preferences and pain points

3. Stay informed:

- Ask ChatGPT to provide summaries of industry news, articles, and research reports, keeping you up-to-date on the latest developments in your niche. Whilst it does not have access to the internet currently you can copy and paste content directly into ChatGPT or use a third party tool that incorporates PDF's and ChatGPT.

By thoroughly researching your market and niche, you can position your ChatGPT-based business for success. In the next section, we'll explore how to develop your business model and identify potential revenue streams.

USEFUL SITES

There are several websites and tools that can provide inspiration, guidance, and resources. Here are some recommendations:

1. Entrepreneur.com (https://www.entrepreneur.com/): A leading source of information and advice for entrepreneurs, featuring articles on business ideas, trends, and best practices.

2. Inc.com (https://www.inc.com/): Offers articles and resources on startups, small businesses, and entrepreneurship, including a section dedicated to business ideas.

3. Business News Daily (https://www.busin

essnewsdaily.com/): Features articles and guides on various aspects of starting a business, including idea generation, business plans, and financing.

4. Bplans (https://www.bplans.com/): Offers free sample business plans, articles, and resources for entrepreneurs to help with idea generation and business plan creation.

5. Reddit (https://www.reddit.com/r/Entrepreneur/): The Entrepreneur subreddit is a community-driven platform where users can share ideas, experiences, and advice on starting and growing businesses.

6. Score.org (https://www.score.org/): A nonprofit organization that provides free mentoring, workshops, and resources for entrepreneurs, including help with business ideas and planning.

7. Small Business Administration (https://www.sba.gov/): Offers resources for entrepre-

neurs, including guides and tools for starting a business, as well as information on government grants and loans.

8. StartupNation (https://startupnation.com/): A community-driven website with articles, podcasts, and forums on various aspects of entrepreneurship, including business ideas and startup advice.

9. Google Trends (https://trends.google.com/trends): This tool allows you to explore trending topics and search queries, which can help you identify potential business opportunities based on current market demands.

10. YouTube Channels: There are many YouTube channels dedicated to entrepreneurship and business ideas, such as Gary Vaynerchuk's channel, and Valuetainment, which offer valuable insights and advice.

11. And don't forget my website www.busines

smaven.io - I also have a YouTube Channel, TikTok & Twitter - if you need more support.

Remember to conduct thorough market research and validate your business idea before investing time and money. It's important to identify a unique selling proposition, understand your target audience, and have a solid plan to launch and grow your business.

CHAT GPT SHORTCUT

You can of course go straight into ChatGPT and ask it to help you define your business idea. Below I have listed the prompt that you can use to get some ideas for a suitable business to start and the reasoning why is included in the output from ChatGPT.

COPY & PASTE PROMPT

BUSINESS IDEAS

"Generate a list of business ideas suitable for someone with the following background, skills, interests, time availability, and startup budget: [describe the person's background, skills, interests, time available to work, and startup budget]. Please provide a brief description of each business idea, its potential income, and why it could be a good fit for the individual, considering their time and financial constraints."

PROMPT EXAMPLE

"Generate a list of business ideas suitable for someone with the following background, skills, interests, time availability, and startup budget: 10 years in social media marketing, expertise in digital design, interests include the gym, dogs and travel, time available to work - just 2 hours a week and startup budget of just $100. Please provide a brief description of each business idea, its potential income, and why it could be a good fit for the individual, considering their time and financial constraints."

CHATGPT RESPONSE

Here are five business ideas that could be suitable for someone with a background in social media marketing, expertise in digital design, and interests in fitness, dogs, and travel, with a limited time availability and a small startup budget:

IDEA 1

Social Media Marketing Consultancy for Gyms: Use your expertise in social media marketing to offer consultancy services to small gyms and fitness studios. You could provide guidance on how to create compelling social media content, design engaging graphics, and increase followers. Potential income could range from $50 to $150 per hour, depending on your experience and the scope of services offered. This business idea would be a good fit for you because it aligns with your interests in fitness, and you can easily leverage your experience in social media marketing.

IDEA 2

Digital Design Services for Dog Products: Leverage your digital design skills to offer customized design services to small businesses that create dog products, such as collars, leashes, and toys. You could offer design services for product packaging, branding, and marketing materials. Potential income could range from $50 to $200 per project, depending on the complexity of the design. This business idea would be a good fit for you because it aligns with your interests in dogs and your expertise in digital design.

IDEA 3

Travel Blog: Use your passion for travel to start a travel blog. You could share your travel experiences, tips, and recommendations with your readers. You could earn income through affiliate marketing, sponsored content, and selling digital products such as travel guides. Potential income could range from $500 to $5,000 per month, depending on the size of your audience and the monetization strategies you use. This business idea would be a good fit for you because it aligns with your interests in travel, and you can easily manage your time commitment.

IDEA 4

Dog Walking Services: Offer dog walking services in your local area. You could target busy pet owners who don't have the time to walk their dogs during the day. Potential income could range from $15 to $25 per hour, depending on the location and the competition in the market. This business idea would be a good fit for you because it aligns with your interests in dogs and your limited time availability.

IDEA 5

Social Media Management for Travel Companies: Use your expertise in social media marketing to manage the social media accounts of small travel companies. You could create engaging content, design graphics, and increase followers. Potential income could range from $50 to $150 per hour, depending on your experience and the scope of services offered. This business idea would be a good fit for you because it aligns with your interests in travel and your experience in social media marketing.

> Overall, these business ideas would be suitable for someone with a background in social media marketing, expertise in digital design, interests in fitness, dogs, and travel, with a limited time availability and a small startup budget. Each business idea has the potential to generate income, and the individual can choose the one that aligns with their interests and time commitment.

With this approach you simply take the idea that most aligns to what you want to do and ask ChatGPT to expand on it. For this example we are going with IDEA 5: Social Media Management for Travel Companies, and I'll show you step-by-step how we set this business up.

"There is no monopoly on becoming a millionaire. If you're jealous of those with more money, don't just sit there and complain - do something to make more money yourself."

Gina Rinehart, Australian billionaire

BUSINESS MODEL AND REVENUE STREAMS

CRAFTING A SUSTAINABLE PROFIT MACHINE

The key to sustainable success in any business lies in the development of a strong business model and the identification of diverse revenue streams. For ChatGPT-based businesses, understanding how to monetize your offerings and generate consistent income is vital. In this section, we will delve into the process of developing your business model, exploring various revenue opportunities, and taking inspiration from successful ChatGPT-based ventures.

A well-crafted business model serves as the backbone of your ChatGPT-based business, outlining how you plan to create, deliver, and capture value. It encompasses everything from your value proposition and target market to your customer acquisition strategy and revenue generation methods. Developing a robust business model requires a deep understanding of your target audience, niche, and competitors, as well as the ability to think critically about how your business will operate and grow.

When it comes to revenue streams, it's essential to think creatively and diversify your income sources. This not only mitigates risks associated with relying on a single revenue stream but also increases your business's overall stability and resilience. By exploring multiple avenues for monetization, you can maximize your revenue potential and capitalize on opportunities within your niche.

In the realm of ChatGPT-based businesses, there are several revenue streams worth considering. For instance, you could generate income through:

1. Subscription-based services:

- Offering premium access to your ChatGPT solution through a monthly or annual subscription fee, providing exclusive features or personalized assistance to subscribers.

2. Pay-per-use or tiered pricing:

- Charging customers based on their usage or access level, enabling you to cater to a wider range of clients with varying needs and budgets.

3. Licensing or white-labeling your technology:

- Allowing other businesses to utilize your ChatGPT solution under their brand, generating revenue through licensing fees or revenue-sharing agreements.

4. Advertising and sponsorship:

- Partnering with businesses or brands that align with your niche, offering advertising space or sponsored content within your ChatGPT platform.

5. Affiliate marketing:

- Promoting third-party products or services related to your niche through your ChatGPT solution, earning a commission for each referral or sale.

6. Customized ChatGPT development:

- Providing bespoke ChatGPT solutions tailored to specific clients or industries, charging a premium for your specialized services.

To truly excel in the business space, it's essential to learn from those who have already found success. Study successful ventures in a similar field, paying close attention to their business models, marketing strategies, and monetization methods. By analyzing their strengths and weaknesses, you can glean valuable insights that will inform your own business strategy.

In conclusion, creating a solid business model and identifying diverse revenue streams are critical components of long-term success for your ChatGPT-based business. By taking the time to understand your target audience, niche, and competitors, you can develop a comprehensive business model that sets the stage for sustainable growth. By exploring various revenue opportunities and learning from successful ChatGPT-based businesses, you can maximize your income potential and establish a thriving enterprise.

Now it's important to note here, that whilst I am talking about a 'ChatGPT based solution' that does not necessarily mean your business needs to incorporate ChatGPT into the service or product you are offering. As an example – a copywriter might have a monthly retainer but use ChatGPT to generate the brainstorming, or could just use ChatGPT to build the business plan as we are doing here. The choice is yours.

WORKSHEET 1:

CREATE A CHATGPT BASED BUSINESS MODEL

Use the following template to develop a comprehensive business model for your ChatGPT- based business:

1. Value Proposition:

- What unique value does your solution offer to your target audience? How does it address their pain points or needs more effectively than competitors?

2. Customer Segments:

- Who are your primary customers, and what

are their characteristics? Are there any secondary customer segments that could benefit from your solution?

3. Customer Relationships:

- How will you establish and maintain relationships with your customers? What channels will you use to communicate with them and provide support

4. Channels:

- How will you reach your target audience and deliver your solution? What marketing and distribution channels will you use to promote and sell your product or service?

5. Key Activities:

- What are the main activities required to create, deliver, and maintain your ChatGPT-based solution? How will you ensure the continuous improvement of your product or service?

6. Key Resources:

- What assets, technologies, or partnerships do you need to execute your business model? How will you acquire and manage these resources?

7. Key Partners:

- Who are your essential partners for delivering your ChatGPT-based solution? What kind of relationships and agreements will you establish with them?

8. Cost Structure:

- What are the primary costs associated with running your ChatGPT-based business? How will you manage and minimize these costs while maintaining quality?

9. Revenue Streams:

- What are the primary sources of income for your ChatGPT-based business? How will you price your products or services, and what payment models will you use?

WORKSHEET 2:

IDENTIFYING AND PRIORITIZING POTENTIAL REVENUE STREAMS

Complete the following worksheet to explore and prioritize potential revenue streams for your ChatGPT-based business:

1. List potential revenue streams:

- Consider various options, such as subscriptions, one-time purchases, advertising, and affiliate marketing

2. Assess the viability of each revenue stream:

- Evaluate the market demand, profitability,

and scalability of each option

3. Prioritize revenue streams:

- Rank the revenue streams based on their potential for success and alignment with your business model

4. Develop a strategy for each prioritized revenue stream:

- Outline how you will implement, market, and manage each revenue stream

CASE STUDIES

Case Study 1:

AI Content Writing Service Business Model:

Offers high-quality, AI- generated content for businesses and individual clients

Revenue Model: Subscription plans for different content volumes and a pay-per-article option for occasional users Success Factors: A strong focus on producing engaging, human-like content and exceptional customer support

Case Study 2:

Chatbot Development Platform Business Model:

Provides a platform for businesses to create, deploy,

and manage ChatGPT-powered chatbots

Revenue Model: Tiered subscription plans based on chatbot features and usage, along with custom pricing for enterprise clients Success Factors: User-friendly interface, extensive customization options, and robust chatbot performance

Case Study 3:

AI-Powered Marketing Analytics Business Model:

Delivers AI-generated insights and recommendations for optimizing marketing campaigns and strategies

Revenue Model: Monthly subscription plans based on the size of the business and the complexity of their marketing needs, as well as a freemium model with limited features for smaller businesses or individuals

Success Factors: In-depth analytics, actionable insights, and seamless integration with popular marketing tools

CHATGPT SHORTCUT

Here's the prompt I use straight into ChatGPT to help me define the business model and revenue streams. Below I have listed the prompt that you can use to get some ideas for a suitable business to start and the reasoning why is included in the output from ChatGPT.

COPY & PASTE PROMPT

"Create a revenue plan for a new business called [Business Name], operating in the [Industry] industry. The business offers [Products/Services] to [Target Market]. The main revenue streams are

[Revenue Streams]. The pricing strategy for the products/services is [Pricing Strategy]. The business aims to achieve [Revenue Goal] in the first year, with a growth rate of [Growth Rate] over the next [Number of Years]. Please consider the following factors in the plan: market research, sales channels, marketing efforts, partnerships, and any other relevant aspects."

PROMPT EXAMPLE

(Using Idea 5 from our Niche Selection Chapter)

"Create a revenue plan for a new business, please provide 10 relevant names we could call this business, operating in the social media industry. The main revenue streams are to be determined by you. The pricing strategy for the products/services is also for you to decide. The business aims to achieve $50,000 in the first year, with a growth rate of whatever you think is possible over the next 5 years. Please consider the following factors in the plan: market research, sales channels, marketing efforts, partnerships, and any other relevant aspects."

CHATGPT'S RESPONSE

Business Name Ideas:

Social Travel Media

Travel Brand Boosters

Social Voyage

Traveler's Media

Social Travel Pro

TravelSocials

Jetsetter Media

Roaming Socials

TravelWave Media

Travel Media Co.

Revenue Plan: Market Research: Conduct market research to identify the target market and the competition. The research should include analyzing the social media presence of small travel companies and their marketing needs.

Sales Channels: The main sales channel will be the company website, which will feature the services and packages offered. Additionally, the company can reach out to potential clients through cold emails, social media outreach, and attending travel industry events.

Marketing Efforts: Develop a comprehensive marketing strategy that includes social media advertising, content marketing, and email marketing. Utilize social media platforms, such as Facebook and Instagram, to showcase successful campaigns and case studies.

Partnerships: Form partnerships with other travel companies and influencers to expand the reach of the business. Collaborate on social media campaigns and promotions to reach a wider audience.

Revenue Streams: Offer customized social media management packages that can include content creation, account management, ad campaigns, and analytics. Additional revenue streams can include offering social media audits, consulting services, and training sessions for small travel companies.

Pricing Strategy: The pricing strategy will be based on the level of service offered and the size of the client's social media presence. The packages can range from $500 to $2,500 per month, depending on the scope of services. The pricing should be competitive with other social media marketing agencies in the travel industry.

Revenue Projections: Based on the market research and pricing strategy, the company aims to achieve $50,000 in the first year. The growth rate for the next five years can be targeted at 50% annually, with revenue projections of $75,000, $112,500, $168,750, $253,125, and $379,688 respectively.

Overall, the key to success in this business is to provide high-quality, customized social media management services to small travel companies. By utilizing various sales channels, developing a comprehensive marketing strategy, forming partnerships, and offering multiple revenue streams, the business can achieve its revenue goals and grow steadily over the next five years.

In this instance we are going to pick "Travel Media Co" as our business name as a temporary solution. We will verify names etc at a later stage.

REMINDER: As a thank you for buying this book. I have created a simple Copy & Paste PDF for you. You can get your copy from the QR code at the back of this book.

"The lack of money is the root of all evil."

Mark Twain, American writer

LAYING THE GROUNDWORK FOR YOUR CHATGPT BUSINESS

ESTABLISHING A COMPREHENSIVE INFRASTRUCTURE

E stablishing a comprehensive infrastructure is a vital component for ensuring the success of your ChatGPT-based venture. In this chapter, we will focus on the key steps you need to take before delving into more specific aspects of your business, such as integration and content creation, which will be covered in the following chapters.

The preliminary steps include:

1. Securing social media accounts

2. Designing your brand

3. Creating your website or app

The Brand

The first essential step is to develop a clear and consistent brand identity. Determine your brand's mission, vision, and values, which will guide the development of your visual and messaging elements. As you create your brand identity, take into account factors such as your target audience, market positioning, and competitive landscape to ensure your brand resonates with your customers. A well-defined brand identity will help you stand out in the marketplace and foster customer loyalty.

Socials

Once you have defined your brand identity, proceed to secure social media accounts on platforms relevant to your target audience. Doing so will allow you to connect with potential customers, share valuable content, and increase your online presence. Remember, consistency across all social media platforms is essential. Use the same username, profile picture, and bio wherever possible to maintain a cohesive brand image. Regularly posting and engaging with your audience will also strengthen your online presence and foster lasting relationships with your customers. Social media platforms such as Facebook, Twitter, Instagram, LinkedIn, and TikTok provide excellent opportunities for promoting your ChatGPT-based business and reaching a wider audience.

With a strong brand identity and social media presence, the next step is to create a website or app that serves as the digital hub for your ChatGPT-based business. Your website or app should be user-friendly, visually appealing, and optimized for both desktop and mobile devices. Prioritize responsive design, as this will ensure your site looks and functions well on various devices and screen sizes.

Website/App

Before you begin building your website or app, research and select a suitable domain name and hosting provider. Choose a domain name that reflects your brand identity and is easy for users to remember. When selecting a hosting provider, consider factors such as server reliability, uptime, and customer support.

Incorporate essential features that cater to your target audience's needs and preferences. These may include a blog, a product or service showcase, customer testimonials, and a contact form. Additionally, integrate ChatGPT into your website or app as a chatbot, virtual assistant, or content generator to enhance user experience and engagement.

As you build your website or app, prioritize search engine optimization (SEO) to improve its visibility in search engine results. Perform keyword research, optimize meta tags, and create high-quality, relevant content that provides value to your audience. This will increase organic traffic, drive leads, and enhance your overall online presence.

In conclusion, laying the groundwork for your ChatGPT business involves crafting a compelling brand identity, securing social media accounts, and creating a website or app that caters to your target audience. By addressing these foundational aspects of your business, you will be well-prepared to integrate ChatGPT and explore the more specific aspects of running a successful AI-powered business in the subsequent chapters of this book. Remember, a well-rounded infrastructure serves as a solid base for your ChatGPT business and will pave the way for long-term success.

WORKSHEET:

LAYING THE GROUNDWORK FOR YOUR CHATGPT BUSINESS

1. Develop a Clear and Consistent Brand Identity

- Determine your brand's mission, vision, and values.
- Consider your target audience, market positioning, and competitive landscape.
- Develop a unique visual and messaging style that represents your brand.

2. Secure Social Media Accounts

- List relevant social media platforms for your target audience.

- Secure usernames, profile pictures, and bios that are consistent with your brand identity.

- Develop a social media posting schedule and content plan.

3. Design and Develop Your Website or App

- Determine the essential features and functionality for your website or app.

- Choose a platform and/or developer to build your website or app.

- Create engaging content and visuals that reflect your brand identity and appeal to your target audience.

- Optimize your website or app for search engines and mobile devices.

4. Set Up Email and CRM Systems

- Choose an email provider and CRM system that fits your business needs (e.g., Gmail, G Suite, Mailchimp, HubSpot).

- Develop email templates and workflows to manage customer communication.

- Segment your audience and develop targeted email campaigns.

5. Implement Marketing Tools and Strategies

- Choose marketing tools to help manage and

analyze your marketing efforts.

- Develop a marketing plan, including goals, objectives, and tactics.

- Explore additional marketing channels, such as paid advertising, content marketing, and influencer partnerships.

6. Establish Strategic Partnerships

- Identify potential partners that align with your brand's values and have a complementary audience.

- Brainstorm partnership ideas, such as webinars, guest blogging, or cross-promotion.

- Reach out to potential partners and establish mutually beneficial collaborations.

7. Develop a Legal and Financial Framework

- Register your business and obtain any necessary licenses and permits.

- Consult with legal and financial experts to ensure compliance with relevant laws and regulations.

- Set up an accounting system and make informed decisions about your business's structure and financial management.

Once you have completed this worksheet, you'll be ready to move on to the more specific aspects of your business, such as ChatGPT integration, which will be covered in the next chapter.

WORKSHEET

BRANDING

Designing Your Brand Identity:

A strong brand identity communicates your unique value proposition and helps you stand out from the competition. Here are some steps to create a memorable and cohesive brand identity:

1. Develop a brand story:

- Craft a compelling narrative that conveys your business's mission, vision, and values. Your brand story should evoke emotions and create a connection between your business and your target audience.

2. Choose a brand name:

- Select a name that is unique, memorable, and relevant to your ChatGPT business. Use tools like Namecheap (https://www.namecheap.com/) or Lean Domain Search (https://leandomainsearch.com/) to find available domain names and brainstorm ideas.

3. Design a logo:

- Create a distinctive logo that represents your brand and resonates with your target audience. You can use online tools like Canva (https://www.canva.com/) or enlist the help of a professional designer through platforms like 99designs (https://99designs.com/) or Dribbble (https://dribbble.com/).

4. Establish a brand color palette and typography:

- Choose colors and fonts that reflect your brand's personality and create a consistent visual identity. Use tools like Coolors (https://coolors.co/) to generate color palettes and Google Fonts (https://fonts.google.com/) to explore font options.

5. Develop brand guidelines:

- Document your brand elements, messaging, and tone of voice to ensure consistency across all channels. This includes logo variations, color codes, font usage, and guidelines for imagery and copywriting.

TOP TIPS

Securing Your Online Presence Establishing a solid online presence is critical for your ChatGPT business's visibility and credibility. Follow these steps to secure your online presence:

1. Register a domain name:

- Choose a domain name that aligns with your brand and is easy to remember. Domain registrars like Namecheap (https://www.namecheap.com/) and GoDaddy (https://www.godaddy.com/) offer domain registration services.

2. Set up professional email addresses:

- Create email addresses using your domain

name to maintain a consistent brand image. Many domain registrars and hosting providers offer email hosting services, or you can use G Suite (https://workspace.google.com/) for a more comprehensive solution.

3. Secure social media handles:

- Register your brand's name on relevant social media platforms to protect your brand identity and expand your reach. Use a tool like KnowEm (https://knowem.com/) to check the availability of your desired username across various platforms.

4. Choose a website hosting provider:

Select a reliable hosting provider that offers the features and support your business needs. Some popular options include SiteGround (https://www.siteground.com/), Bluehost (https://www.blueho

st.com/), and WP Engine (https://wpengine.com/) for WordPress-based websites.

Website

Developing Your Website or App A well-designed website or app is crucial for showcasing your ChatGPT-based offerings and attracting customers. Here's a guide to building your platform:

1.Determine your site's purpose and structure:

- Outline the primary goals of your website or app and create a sitemap to organize your content. Identify the key pages and features your platform needs to effectively serve your target audience.

2. Select a content management system (CMS) or app development platform:

- Choose a CMS like WordPress (https://wordpress.org/), Joomla (https://www.joomla.org/), or Drupal (https://www.drupal.org/) for building your website, or consider using a website builder like Wix (https://www.wix.com/) or Squarespace (https://www.squarespace.com/) for a more streamlined approach. For app development, consider using platforms like Flutter (https://flutter.dev/) or React Native (https://reactnative.dev/).

3. Optimize your content:

- Ensure your website or app content is engaging, informative, and optimized for search engines. Use keyword research tools to identify relevant keywords and incorporate them into your content. Create high-quality, unique, and value-driven content that addresses the needs and interests of your target audience.Set up analytics: Install tracking tools like Google Analytics (https://analytics.google.com/) or Hotjar (https://www.hotjar.com/) to monitor your website or app's performance and gather insights into user behavior. Use this data to optimize your platform, improve user experience, and drive business growth.

4. Design your website or app:

- Create a visually appealing and user-friendly interface that aligns with your brand identity. Focus on intuitive navigation, mobile responsiveness, and fast-loading pages. You can use pre-built themes and templates or hire a professional designer and developer to create a custom solution.

By following these steps, you'll establish a strong foundation for your ChatGPT business, paving the way for success as you move forward with marketing, promotion, and scaling your enterprise.

WORKSHEET

LAYING THE GROUNDWORK FOR YOU CHATGPT BUSINESS

1. Develop a Clear and Consistent Brand Identity

- Determine your brand's mission, vision, and values.

- Consider your target audience, market positioning, and competitive landscape.

- Develop a unique visual and messaging style that represents your brand.

2. Secure Social Media Accounts

- List relevant social media platforms for your target audience.

- Secure usernames, profile pictures, and bios that are consistent with your brand identity.

- Develop a social media posting schedule and content plan.

3. Design and Develop Your Website or App

- Determine the essential features and functionality for your website or app.

- Choose a platform and/or developer to build your website or app.

- Create engaging content and visuals that reflect your brand identity and appeal to your target audience.

- Optimize your website or app for search engines and mobile devices.

4. Set Up Email and CRM Systems

- Choose an email provider and CRM system that fits your business needs (e.g., Gmail, G Suite, Mailchimp, HubSpot).

- Develop email templates and workflows to manage customer communication.

- Segment your audience and develop targeted email campaigns.

5. Implement Marketing Tools and Strategies (Strategies we will cover fully in a few chapters)

- Choose marketing tools to help manage and analyze your marketing efforts.

- Develop a marketing plan, including goals, objectives, and tactics.

- Explore additional marketing channels, such

as paid advertising, content marketing, and influencer partnerships.

6. Establish Strategic Partnerships

- Identify potential partners that align with your brand's values and have a complementary audience.

- Brainstorm partnership ideas, such as webinars, guest blogging, or cross-promotion.

- Reach out to potential partners and establish mutually beneficial collaborations.

7. Develop a Legal and Financial Framework

- Register your business and obtain any necessary licenses and permits.

- Consult with legal and financial experts to ensure compliance with relevant laws and regulations.

- Set up an accounting system and make informed decisions about your business's structure and financial management.

Once you have completed this worksheet, you'll be ready to move on to the more specific aspects of your business, such as ChatGPT integration, which will be covered in the next chapter.

CHATGPT SHORTCUT

"Can you help me create a comprehensive brand identity for my [Business] targeting [Audience], including a mission statement, vision statement, target audience profile, market positioning, competitive landscape analysis with reasoning, and suggestions for a unique visual and messaging style and color palette?"

PROMPT EXAMPLE

"Can you help me create a comprehensive brand identity for my social media marketing company targeting small travel agencies, including a mission statement, vision statement, target audience profile, market positioning, competitive landscape analysis with reasoning, and suggestions for a unique visual and messaging style and color palette?"

CHATGPT'S RESPONSE

Yes, I can definitely help you with that. Here's a comprehensive guide to creating a brand identity for your social media marketing company, Travel Media Co, targeting small travel agencies, including a competitive landscape analysis with reasoning:

Mission statement: Our mission is to provide effective social media marketing solutions that help small travel agencies increase their online visibility, engagement, and bookings, while delivering exceptional customer service and results.

Vision statement: Our vision is to become the leading provider of social media marketing services for small travel agencies, recognized for our innovative solutions, industry expertise, and commitment to customer satisfaction.

Target audience profile: Our target audience is small travel agencies that specialize in adventure, cultural, and eco-tourism, with annual revenues between $500,000 and $2 million. They are typically located in North America and Europe and are looking to increase their online presence, reach new customers, and drive more bookings through social media marketing.

Market positioning: We position ourselves as a premium social media marketing company that offers customized solutions tailored to the unique needs of small travel agencies. We differentiate ourselves from competitors by combining industry expertise with cutting-edge technology and personalized customer service.

Competitive landscape analysis with reasoning: Our primary competitors include other social media marketing agencies that serve small travel agencies, such as Socialight Marketing, Travel Marketing Solutions, and Outbound Communications.

Socialight Marketing has a strong reputation in the market and offers a variety of services, including social media marketing, SEO, and content marketing. However, they tend to take a one-size-fits-all approach to their solutions and may not provide the same level of personalized service as Travel Media Co.

Travel Marketing Solutions specializes in providing digital marketing solutions for the travel industry and has a similar target audience to Travel Media Co. However, they tend to focus more on larger clients and may not have the same level of expertise in social media marketing specifically.

Outbound Communications is a smaller agency that focuses on outbound marketing and lead generation. While they may not be direct competitors, they do offer some similar services and may be a viable alternative for small travel agencies that prioritize lead generation over online visibility and engagement.

Suggestions for a unique visual and messaging style and color palette: Our visual and messaging style is designed to convey our brand personality, which is friendly, professional, and approachable. Our color palette features shades of blue and green, representing travel, nature, and adventure. We use high-quality imagery that showcases the beauty and diversity of travel destinations, combined with clear and concise messaging that communicates our value proposition and expertise.

I hope this provides you with a better understanding of how to create a comprehensive brand identity for your social media marketing company, Travel Media Co, targeting small travel agencies, with a focus on competitive landscape analysis and reasoning. Let me know if you have any further questions or need any additional assistance.

Now it's essential to note that ChatGPT was trained on data up until 2021 – so this information is likely to be out of date. So I highly recommend you do your own research in Google to double check.

REMINDER: There is a QR code at the back of this book that gives you all the Copy and Paste Prompts in a convenient PDF

"To succeed, jump as quickly at opportunities as you do at conclusions."

Benjamin Franklin, One of the Founding Fathers of the United States

CHATGPT INTEGRATION AND OPTIMIZATION

MAXIMIZING EFFICIENCY AND USER EXPERIENCE

Incorporating ChatGPT into your business demands meticulous planning and optimization to guarantee a smooth experience for your users and cost-efficiency for your operations. In this section, we will outline the essentials for planning your ChatGPT integration process, provide guidance for selecting the most suitable ChatGPT API plan, and share valuable tips to optimize performance without compromising your budget. Now it's important to remember here that you do not necessarily need to integrate ChatGPT into your business beyond teh business plan – this step is entirely dependent on your needs.

A successful ChatGPT integration hinges on a well-structured plan that addresses your business's unique needs and objectives. This plan should encompass all aspects of implementation, from understanding the technical requirements to tailoring the ChatGPT solution to fit your specific niche. By devising a comprehensive integration strategy, you can mitigate potential challenges and ensure that your ChatGPT-based business is poised for success.

The first step in this process is to evaluate your business's current infrastructure and identify areas where ChatGPT can add value. Consider how ChatGPT can enhance your customer experience, streamline internal processes, or generate new revenue streams. This assessment will help you prioritize your integration efforts and allocate resources effectively.

Next, familiarize yourself with the technical requirements of integrating ChatGPT into your business. This includes understanding the API documentation, identifying the necessary software and hardware components, and determining any additional development or customization needs. By gaining a solid grasp of these technical aspects, you can develop a realistic integration timeline and budget.

Selecting the right ChatGPT API plan is critical for both performance and cost-effectiveness. To make an informed decision, consider the following factors:

Usage requirements:

Assess your business's anticipated usage levels, including the number of tokens, requests, and concurrent users you expect to handle. Choose a plan that accommodates these needs while also providing room for growth.

Budget constraints:

Evaluate the costs associated with each API plan, taking into account not only the subscription fees but also any additional expenses related to development, customization, or maintenance.

Support and service level agreements (SLAs):

Compare the support and SLAs offered by each API plan, ensuring that your chosen plan provides the level of assistance and reliability your business requires.

Finally, optimizing your ChatGPT integration is essential for maximizing performance and minimizing costs. Some key optimization strategies include:

Fine-tuning your ChatGPT model:

- Adjust the parameters of your model to

strike the optimal balance between response quality and token usage, reducing unnecessary expenses without compromising the user experience.

Implementing caching and rate limiting:

- Utilize caching to store and retrieve frequently used responses, and apply rate limiting to control the number of requests per user, ensuring efficient resource allocation and preventing API overuse.

Monitoring usage and performance:

- Regularly analyze your ChatGPT usage and performance data, identifying areas for improvement and making data-driven adjustments to your integration strategy.

In conclusion, integrating and optimizing ChatGPT for your business is a complex but essential process that requires careful planning, evaluation, and ongoing refinement. By developing a comprehensive integration strategy, selecting the most suitable API plan, and continually optimizing your ChatGPT solution, you can create a seamless, cost-effective experience for your users and establish a strong foundation for your ChatGPT-based business.

WORKSHEET:

PLANNING YOUR CHATGPT INTEGRATION PROCESS

Use the following template to map out your ChatGPT integration process:

1.Define your objectives:

- What specific goals do you want to achieve with ChatGPT integration?

- How will ChatGPT contribute to your business's overall success?

2. Identify integration points:

- Where in your business processes will Chat-

GPT be integrated?

- What tasks or functions will it perform?

3. Determine technical requirements:

- What technical resources, such as infrastructure, software, or hardware, are needed for ChatGPT integration?
- How will you ensure compatibility with your existing systems?

4. Develop a timeline:

- Outline the key milestones and deadlines for your ChatGPT integration process
- Allocate sufficient time for testing, optimization, and troubleshooting

5. Assign roles and responsibilities:

- Identify team members responsible for

implementing, managing, and maintaining your ChatGPT integration

- Define their specific roles, tasks, and responsibilities

6. Plan for ongoing maintenance and optimization:

- Outline a strategy for monitoring ChatGPT's performance and making necessary adjustments

- Establish a process for incorporating updates, improvements, and new features

CHECKLIST:

Evaluating and Selecting the Right ChatGPT API Plan

Consider the following factors when choosing a ChatGPT API plan that best suits your business needs:

1. API rate limits:

- Assess the number of requests per minute (RPM) and tokens per minute (TPM) allowed by each plan
- Ensure the selected plan supports your business's expected usage volume

2. Pricing:

- Compare the cost of different plans and evaluate their value based on your expected usage

- Factor in additional costs, such as overage fees, if you exceed the plan's limits

3. Support:

- Assess the level of customer support provided with each plan

- Consider whether priority support is necessary for your business

4. Features and functionality:

- Review the features and capabilities offered by each plan

- Ensure the selected plan includes the features crucial to your business's success

TOP TIPS:

OPTIMIZING CHATGPT PERFORMANCE AND MINIMIZING COSTS

1. Fine-tune prompts and settings:

- Experiment with different prompts and settings to achieve the desired output quality and reduce the number of tokens used in each request

2. Batch requests:

- Combine multiple requests into a single API call to improve efficiency and reduce the number of API calls

3. Monitor usage:

- Keep track of your API usage to identify trends, patterns, and potential bottlenecks

- Adjust your usage patterns and API plan as needed to optimize cost-effectiveness

4. Cache results:

- Store frequently requested responses in a cache to minimize redundant API calls and reduce costs

By following these steps and carefully planning your ChatGPT integration and optimization, you'll be well on your way to leveraging the power of ChatGPT effectively and efficiently in your business. In the next section, we'll explore content creation and automation using ChatGPT.

CHATGPT SHORTCUT

Here's the prompt I would use to help to understand where you could use ChatGPT and automation in future to streamline your business. You obviously don't need to implement all the recommendations immediately, but creating a business that operates with as little human interference with mechanical or admin tasks is probably a wise move.

Below I have listed the prompt that you can use to get some ideas for streamlining your operations with ChatGPT in your new business.

COPY & PASTE PROMPT

"What are some areas in [insert type of business] where ChatGPT or automation can add value and improve efficiency? Bearing in mind this is a new business, with a limited budget of [budget] and only [weekly_hours] hours available a week. Can you create a plan and timeline for ChatGPT and automation to be added to the business, the most important first? If you recommend any tools, please provide links and reasons why you chose that specific tool in the plan."

PROMPT EXAMPLE

(Using the example Travel Media Co)

"What are some areas in social media management business where ChatGPT or automation can add value and improve efficiency? Bearing in mind this is a new business, with a limited budget of

$100 and only 2 hours available a week. Can you create a plan and timeline for ChatGPT and automation to be added to the business, the most important first. If you recommend any tools, please provide links and reasons why you chose that specific tool in the plan."

CHATGPT'S RESPONSE

Considering the limited budget and time availability for a new social media management business, it's essential to prioritize the integration of ChatGPT and automation in areas that provide the most significant immediate impact. Here's a plan, including recommended tools and their benefits:

1. Content Creation (Weeks 1-2):

- Use ChatGPT to generate engaging text-based content such as social media posts and captions.

- Utilize the free trial of OpenAI's API, if available, for this purpose: https://beta.openai.com/signup/ (It's the source of ChatGPT and provides direct access to the model.)

2. Content Curation (Weeks 3-4):

- Use a free content curation tool like Feedly (https://feedly.com) to identify relevant content for sharing on social media platforms.

- Feedly is a popular RSS reader that allows you to follow websites, blogs, and publications in one place, making it easier to find shareable content.

3. Social Media Scheduling (Weeks 5-6):

- Use a free social media scheduling tool like Buffer (https://buffer.com) to automate posting at optimal times.

- Buffer offers a free plan that allows you to manage 3 social accounts and schedule up to 10 posts per account. It's user-friendly and widely used in the industry.

4. Basic Sentiment Analysis (Weeks 7-8):

- Leverage a free sentiment analysis tool like RapidAPI's Sentiment Analysis API (https://rapidapi.com/twinword/api/sentiment-analysis1/) to gauge user sentiment and the success of your campaigns.

- This API allows you to analyze text to understand the sentiment behind it. RapidAPI offers a free plan with limited usage, which should be suitable for a small business.

5. Simple Chatbot for Customer Support (Weeks 9-10):

- Create a basic chatbot using Tars (https://tars.io) to handle routine customer inquiries.

- Tars is a chatbot builder platform that offers a free trial. While it doesn't directly use ChatGPT, you can leverage ChatGPT-generated content to create engaging chatbot scripts, given your budget constraints.

6. Performance Analysis and Reporting (Weeks 11-12):

- Use the free analytics tools provided by social media platforms (Facebook Insights, Twitter Analytics, etc.) to track performance and generate reports.

- These native tools offer in-depth insights into your social media performance and are already integrated into the platforms.

7. Community Management (Ongoing):

- Manually engage with the community and respond to comments and messages when possible, given the 2-hour weekly availability.

8. Continuous Monitoring and Optimization (Ongoing):

- Regularly analyze performance data and make adjustments to your content, posting times, and tool usage to maximize efficiency within your limited budget.

Please note that this plan assumes you can find suitable low-cost or free tools for the specific tasks mentioned. The timeline is also condensed to fit within a

12-week period, considering the limited weekly time investment. Adjust the plan as necessary to fit your unique requirements and resources.

Remember this is just a template for you. If you need to tweak the plan, then do so! You don't have to rely solely on what ChatGPT tells you. This is a tool for you, albeit an excellent time saving and highly informative tool.

> "Money won't create success, the freedom to make it will."

Nelson Mandela, First President of South Afric

CONTENT CREATION AND AUTOMATION

STREAMLINING ENGAGING INTERACTIONS

Content plays a pivotal role in the success of any ChatGPT-based business, and mastering the art of creating and automating content is crucial for sustained growth. In this section, we will provide a framework for designing a content strategy tailored to your ChatGPT-based venture, offer guidance for planning and scheduling content creation tasks, and share valuable tips for leveraging ChatGPT to generate high-quality, engaging content.

Content Strategy

A well-crafted content strategy serves as a roadmap for your ChatGPT-based business, guiding your content creation efforts and ensuring that your messaging aligns with your target audience's needs and preferences. By developing a content strategy that is customized to your niche and audience, you can create a consistent, impactful experience for your users, strengthening your brand identity and fostering customer loyalty.

The first step in crafting your content strategy is to identify the goals and objectives that your content will serve. This may include educating your audience about your niche, providing solutions to their pain points, or showcasing the benefits of your ChatGPT-based services. By establishing clear content goals, you can better align your content with your business objectives and measure the effectiveness of your efforts.

Channels

Next, consider the various content formats and channels that are best suited to your target audience and niche. This may include blog posts, videos, social media updates, podcasts, or email newsletters. By selecting formats and channels that resonate with your audience, you can maximize the reach and impact of your content.

To streamline your content creation process, develop a content calendar that outlines the topics, formats, and channels for each piece of content, as well as the associated deadlines and responsibilities. This calendar will serve as a valuable tool for planning and scheduling content creation tasks, ensuring that your content remains consistent and relevant to your audience.

Utilizing ChatGPT to produce high-quality, engaging content offers several benefits, including reduced content creation time, increased efficiency, and the ability to generate a diverse range of content ideas. Some tips for leveraging ChatGPT in your content creation process include:

Idea generation:

- Use ChatGPT to brainstorm content topics related to your niche, helping you uncover areas of interest to your target audience and keep your content fresh and varied.

Drafting and editing:

- Employ ChatGPT to draft content based on your chosen topics, and then use it to refine and polish your drafts, ensuring that your content is well-structured, informative, and engaging.

Personalization and adaptation:

- Customize ChatGPT-generated content to address the specific needs and preferences of your audience, adapting the tone, style, and format to create a tailored experience for your users.

Automation and scaling:

- Implement ChatGPT to automate content creation tasks, allowing you to scale your content efforts and maintain a consistent output without sacrificing quality or increasing your workload.

In conclusion, content creation and automation are essential components of a successful ChatGPT-based business. By developing a tailored content strategy, planning and scheduling content creation tasks, and harnessing the power of ChatGPT to produce high-quality, engaging content, you can build a strong foundation for your business and foster long-term growth.

WORSHEET 1:

CRAFTING A CONTENT STRATEGY FOR YOUR CHATGPT-BASED BUSINESS

Use this template to develop a content strategy that aligns with your business goals and target audience:

1.Set content goals:

- What do you aim to achieve with your content (e.g., brand awareness, lead generation, customer engagement)?

- How will your content goals support your overall business objectives?

2. Identify your target audience:

- Who are you creating content for, and what are their needs, preferences, and pain points?

- How will your content address these needs and preferences?

3. Determine content types and formats:

- What types of content (e.g., blog posts, social media updates, email newsletters, videos) will resonate with your audience and achieve your content goals?

- How will you present your content in engaging and visually appealing formats?

4. Plan your content mix:

- Outline a diverse mix of content topics and formats to maintain audience interest and engagement

- Consider incorporating evergreen content,

trending topics, and user-generated content

5. Establish a content calendar:

- Schedule your content creation and distribution tasks to ensure a consistent flow of fresh, relevant content

- Allow for flexibility to accommodate time-sensitive content and emerging trends

WORKSHEET 2:

PLANNING AND SCHEDULING CONTENT CREATION TASKS

Complete the following worksheet to plan and schedule your content creation tasks:

1. List content topics and formats:

- Brainstorm a variety of content ideas that align with your content strategy and audience preferences

2. Assign content creation tasks:

- Delegate content creation responsibilities to team members or freelancers, specifying the required formats and deadlines

3. Schedule content creation deadlines:

- Set realistic deadlines for completing each content creation task, allowing sufficient time for editing and revisions

4. Plan content distribution:

- Determine the appropriate channels and distribution methods for each content piece, ensuring maximum reach and engagement

TOP TIPS:

UTILIZING CHATGPT FOR PRODUCING HIGH-QUALITY, ENGAGING CONTENT

1. Generate content ideas:

- Use ChatGPT to brainstorm content topics and angles, ensuring you always have fresh ideas to explore

2. Create drafts and outlines:

- Leverage ChatGPT to generate drafts or outlines for your content pieces, streamlining the content creation process

3. Enhance readability:

- Ask ChatGPT to simplify complex concepts or rephrase sentences for better readability and engagement

4. Optimize content for SEO:

- Utilize ChatGPT to incorporate relevant keywords and generate SEO-friendly headlines, meta descriptions, and image alt text

5. Edit and proofread:

- Use ChatGPT to identify grammar, spelling, and punctuation errors, ensuring your content is polished and professional

By implementing a content strategy and effectively utilizing ChatGPT for content creation and automation, your ChatGPT-based business will benefit from increased visibility, engagement, and credibility. In the next section, we'll discuss marketing and promotion strategies for your ChatGPT-based business.

CHATGPT SHORTCUT

Here's the prompt I would use to help to understand where you could use ChatGPT to help define a content plan for your business. You will need to determine exactly what the steps will entail depending on your business but this is an excellent foundation to kickstart your plan.

Below I have listed the prompt that you can use to get some ideas for your content plan with ChatGPT in your new business.

COPY AND PASTE PROMPT

"Using the following revenue plan: [copy and paste the revenue plan generated by ChatGPT in Chapter 4] Can you create a comprehensive social media content strategy for my business? Please provide guidance on how to set content goals, identify my target audience, determine the most effective content types and formats, plan a diverse content mix, and establish a content calendar. Once you have a thorough understanding of my business goals and target audience, please outline a content strategy that includes specific recommendations for content types, formats, topics, and distribution channels, as well as a content calendar that includes timelines and deadlines."

PROMPT EXAMPLE

(Using the example Travel Media Co)

Using the following revenue plan: "Revenue Plan: Market Research: Conduct market research to identify the target market and the competition. The research should include analyzing the social media presence of small travel companies and their marketing needs. Sales Channels: The main sales channel will be the company website, which will feature the services and packages offered. Additionally, the company can reach out to potential clients through cold emails, social media outreach, and attending travel industry events. Marketing Efforts: Develop a comprehensive marketing strategy that includes social media advertising, content marketing, and email marketing. Utilize social media platforms, such as Facebook and Instagram, to showcase successful campaigns and case studies. Partnerships: Form partnerships with other travel companies and influencers to expand the reach of the business. Collaborate on social media campaigns and promotions to reach a wider audience. Revenue Streams: Offer customized social media management packages that can include content creation, account management, ad campaigns, and analytics. Additional revenue streams can include offering social media audits, consulting services, and training sessions for small travel companies. Pricing Strategy: The pricing strategy will be based on the level of service offered and the size of the client's social media presence. The packages can range from $500 to $2,500 per month, depending on the scope of services. The pricing should be competitive with other social media marketing agencies in the travel industry. Revenue Projections: Based on the market research and pricing strategy, the company aims to achieve $50,000 in the first year. The growth rate for the next five years can be targeted at 50% annually, with revenue projections of $75,000, $112,500, $168,750, $253,125, and $379,688 respectively. Overall, the key to success in this business is to provide high-quality, customized social media management services to small travel companies. By utilizing various sales channels, developing a comprehensive marketing strategy, forming partnerships, and offering multiple revenue streams, the business can achieve its revenue goals and grow steadily over the next five years." can you create a comprehensive social media content strategy for my business? Please provide guidance on how to set content goals, identify my target audience, determine the most effective content types and formats, plan a diverse content mix, and establish a content calendar. Once you have a thorough understanding of my business goals and target audience, please outline a content strategy that includes specific recommendations for content types, formats, topics, and distribution channels, as well as a content calendar that includes timelines and deadlines."

CHATGPT'S RESPONSE

Sure, I can help you develop a social media content strategy for your business based on the revenue plan you provided. Here's an outline of the strategy:

1. Set Content Goals:

- Increase brand awareness and social media engagement among small travel companies
- Drive website traffic and lead generation

through social media

2. Identify Your Target Audience:

- Small travel companies with a social media presence
- Potential clients who are attending travel industry events and trade shows

3. Determine Content Types and Formats:

- Blog posts highlighting successful campaigns and case studies
- Social media updates featuring visually appealing images and videos
- Email newsletters providing social media tips and industry news

4. Plan Your Content Mix:

- Develop a diverse mix of content topics and formats to maintain audience interest and engagement

- Incorporate evergreen content, trending topics, and user-generated content

- Collaborate with travel influencers and partner companies on social media campaigns and promotions

5. Establish a Content Calendar:

- Schedule your content creation and distribution tasks to ensure a consistent flow of fresh, relevant content

- Allow for flexibility to accommodate time-sensitive content and emerging trends

6. Distribution Channels:

- Promote content on social media platforms such as Facebook and Instagram, utilizing targeted advertising to reach the desired audience

- Send email newsletters to current and potential clients to keep them informed about the latest social media trends and industry news

- Collaborate with partner companies and influencers to reach a wider audience

7. Visual and Messaging Style:

- Use visually appealing images and videos that showcase travel destinations and experiences

- Utilize a messaging style that emphasizes the importance of social media for small travel companies and the benefits of customized social media management services

8. Color Palette:

- Utilize a color palette that incorporates shades of blue and green to represent travel and the natural beauty of destinations

Overall, by implementing this social media content strategy, you can achieve your business goals of increasing brand awareness, driving website traffic, and generating leads among small travel companies.

This will obviously need work on your side, but is a great starting point for you to build from.

REMINDER: The Copy & Paste version of this prompt is available in a handy pdf file which you can download with the QR code at the back of this book.

"Some people call this artificial intelligence, but the reality is this technology will enhance us. So instead of artificial intelligence, I think we'll augment our intelligence."

Ginni Rometty, Former President & CEO
IBM

MARKETING AND PROMOTION

ACCELERATING YOUR BUSINESS GROWTH

A skillfully executed marketing and promotion plan is essential for propelling growth and ensuring the success of your ChatGPT-based venture. In this section, we will walk you through the development of a marketing plan customized to your business, offer guidance for evaluating and selecting the most effective marketing channels, and share insights on leveraging ChatGPT to boost your marketing efforts.

A comprehensive marketing plan serves as a blueprint for your ChatGPT-based business's promotional activities, outlining your marketing goals, target audience, messaging, and strategies. By crafting a marketing plan that is tailored to your niche and audience, you can maximize the impact of your marketing efforts and build a strong, recognizable brand.

Marketing Plan

The first step in creating your marketing plan is to establish clear, measurable goals that align with your overall business objectives. These goals might include increasing brand awareness, driving customer acquisition, or boosting customer engagement. By setting specific, achievable targets, you can monitor the effectiveness of your marketing activities and make data-driven adjustments to your strategy.

Next, input your target audience information including their needs, preferences, and behaviors.

This includes the demographics, interests, pain points, and preferred communication channels. Armed with this knowledge, you can tailor your marketing messaging and tactics to resonate with your audience and drive desired outcomes.

Evaluating and selecting the right marketing channels is crucial for the success of your marketing plan. To make an informed decision, consider the following factors:

Audience reach and engagement:

- Assess the potential reach and engagement of each channel, taking into account your target audience's preferences and habits.

Cost-effectiveness:

- Compare the costs associated with each channel, including ad spend, content creation, and management expenses. Select channels that offer the best return on investment for your budget.

Alignment with your brand and goals:

- Choose channels that align with your brand identity and marketing goals, ensuring that your promotional efforts are consistent and cohesive.

Enhanced Marketing

Harnessing the power of ChatGPT can significantly enhance your marketing efforts by streamlining content creation, personalizing messaging, and enabling innovative promotional tactics. Some tips for utilizing ChatGPT in your marketing and promotion plan include:

Content generation:

- Leverage ChatGPT to generate marketing content tailored to your audience, such as blog posts, social media updates, email campaigns, and ad copy.

Personalized messaging:

- Use ChatGPT to create personalized marketing messages based on user data and preferences, fostering deeper connections with your audience and improving conversion rates.

Customer support and engagement:

- Implement ChatGPT to provide real-time customer support and interactive experiences, such as live chat, chatbots, or virtual assistants, enhancing customer satisfaction and loyalty.

Creative campaigns:

- Employ ChatGPT to brainstorm innovative marketing campaign ideas, helping you stand out from the competition and capture your audience's attention.

In conclusion, a well-executed marketing and promotion plan is vital for the growth and success of your ChatGPT-based business. By developing a customized marketing plan, evaluating and selecting the most effective channels, and harnessing the capabilities of ChatGPT, you can maximize the impact of your promotional efforts and build a thriving, recognizable brand.

TEMPLATE:

DEVELOPING A MARKETING PLAN

Use this template to create a comprehensive marketing plan that supports your business goals:

1. Set marketing objectives:

- What do you aim to achieve with your marketing efforts (e.g., brand awareness, lead generation, customer acquisition)?
- How will your marketing objectives align with your overall business goals?

2. Define your target audience:

- Who are your ideal customers, and what are their characteristics, preferences, and pain points?

- How will your marketing efforts resonate with and address the needs of this audience?

3. Identify marketing channels:

- Which marketing channels (e.g., social media, email, content marketing, paid advertising) will be most effective for reaching your target audience and achieving your objectives?

- How will you allocate your marketing budget and resources across these channels?

4. Develop marketing messages and creative assets:

- What key messages will resonate with your audience and motivate them to take action?

- How will you create compelling visuals, copy, and other assets to support your marketing efforts?

5. Plan and schedule marketing campaigns:

- Outline your marketing campaigns, including objectives, target audience, channels, messages, assets, and timeline

- Ensure a consistent flow of marketing initiatives to maintain audience interest and engagement

6. Measure and analyze performance:

- What key performance indicators (KPIs) will you track to evaluate the success of your marketing efforts?

- How will you use data and insights to inform future marketing decisions and optimize your strategy?

CHECKLIST:

EVALUATING AND SELECTING MAKETING CHANNELS

Consider the following factors when choosing the most effective marketing channels for your ChatGPT-based business:

1.Target audience preferences:

- Where does your target audience spend time online, and which channels do they prefer for consuming content and engaging with brands?

2. Alignment with objectives:

- Which channels are best suited for achieving

your specific marketing objectives, such as brand awareness, lead generation, or customer acquisition?

3. Cost and resource requirements:

- What are the financial and human resource investments required for each channel, and how do these align with your budget and capacity?

4. Scalability and adaptability:

- Can the chosen channels be scaled up or down as your business grows, and can they adapt to changing market conditions and trends?

5. Track record and case studies:

- What is the proven success of each channel in delivering results for similar businesses or industries?

TOP TIPS:

USINES CHATGPT TO ENHANCE MARKETING EFFORTS

1. Generate engaging copy:

- Use ChatGPT to create compelling headlines, ad copy, email subject lines, and social media captions that grab your audience's attention

2. Personalize marketing messages:

- Leverage ChatGPT to craft personalized messages tailored to different audience segments, improving engagement and conversion rates

3. Optimize content for SEO:

- Utilize ChatGPT to generate keyword-rich content, meta descriptions, and image alt text that improve your search engine visibility

4. Streamline content creation:

- Speed up the content creation process by using ChatGPT to generate initial drafts, outlines, and ideas for blog posts, articles, and other marketing materials

5. Monitor and analyze social media sentiment:

Employ ChatGPT to analyze social media conversations, identify trends and sentiment, and inform your marketing strategy

By developing a marketing plan tailored to your ChatGPT-based business and effectively harnessing the power of ChatGPT in your marketing efforts, you'll be well-equipped to reach your target audience, achieve your objectives, and drive growth for your business. In the next section, we'll discuss strategies for enhancing customer support and retention using ChatGPT.

CHATGPT SHORTCUT

Here's the prompt I would use to help to understand where you could use ChatGPT to create a marketing and promotion plan for your business. You will need to determine exactly what the steps will entail depending on your business but this is an excellent foundation to kickstart your plan.

Below I have listed the prompt that you can use to get some ideas for the outline to your marketing plan with ChatGPT in your new business.

COPY AND PASTE PROMPT

"Create a full marketing plan for a new [industry] company called [company name] that targets [target audience]. Please be conscious that we are starting with [budget]and only have [time available to work on the business]. The plan should include market research, sales channels, marketing efforts, partnerships, revenue streams, pricing strategy, revenue projections, and a social media content strategy. Please give as much detail as possible, including any relevant urls and examples to assist us. Additionally, suggest ways in which ChatGPT can assist in the automation of tasks within the marketing plan."

PROMPT EXAMPLE

(USING THE TRAVEL MEDIA CO)

"Create a full marketing plan for a new

social media marketing company called Travel Media Co that targets small travel agencies. Please be conscious that we are starting with $100 budget and only have 2 hours of work available to work on this plan. The plan should include market research, sales channels, marketing efforts, partnerships, revenue streams, pricing strategy, revenue projections, and a social media content strategy. Please give as much detail as possible, including any relevant urls and examples to assist us. Additionally, suggest ways in which ChatGPT can assist in the automation of tasks within the marketing plan."

CHATGPT'S RESPONSE

Marketing Plan for Travel Media Co

1. Market Research:

a. Target Audience:

- Small travel agencies who need social media marketing services to boost their online presence.
- Travel-related businesses like hotels, hostels, and tour operators.

b. Competitor Analysis:

- Identify key competitors and study their services, pricing, and marketing strategies.
- Use free online tools like SimilarWeb (https://www.similarweb.com/) and SEMrush (https://www.semrush.com/) for insights on competitor websites.

c. Industry Trends:

- Research the latest trends in the travel industry and social media marketing.

- Use Google Trends (https://trends.google.com/trends/) to find popular search terms and topics.

- Follow travel and marketing blogs, news websites, and influencers to stay updated.

2. Sales Channels:

a. Website:

- Create a simple, professional website using free website builders like Wix (https://www.wix.com/) or WordPress (https://wordpress.com/).

- Showcase your services, pricing, portfolio, and testimonials.

b. Social Media:

- Create profiles on platforms like Facebook, Instagram, and LinkedIn.

- Post regularly and engage with your audience to build a strong online presence.

3. Marketing Efforts:

a. Content Marketing:

- Write informative blog posts on topics related to travel and social media marketing.
- Share your content on social media and online forums to drive traffic.

b. Email Marketing:

- Collect email addresses of potential clients and send them newsletters and promotions.
- Use free email marketing tools like Mailchimp (https://mailchimp.com/) or Sendinblue (https://www.sendinblue.com/).

c. Networking:

- Attend local events and meetups related to

travel and marketing.

- Network with travel agency owners and other professionals in the industry.

4. Partnerships:

a. Collaborate with complementary businesses like travel bloggers, photographers, and videographers.
b. Offer discounts or commission for referrals from existing clients.

5. Revenue Streams:

a. Social Media Management:

- Offer social media management services for a monthly fee.

- Provide content creation, scheduling, and analytics reporting.

b. Social Media Advertising:

- Help clients run paid advertising campaigns on platforms like Facebook and Instagram.

6. Pricing Strategy:

a. Research competitor pricing and create a competitive pricing structure. b. Offer tiered pricing based on the level of service and customization required by clients.

7. Revenue Projections:

- Estimate the number of clients and average monthly revenue for the first year.

- Adjust your projections based on your initial success and growth.

8. Social Media Content Strategy:

a. Curate content that is relevant, engaging, and visually appealing. b. Mix promotional content with educational, inspirational, and entertaining content. c. Create a content calendar and schedule posts in advance using free tools like Hootsuite (https://hootsuite.com/) or Buffer (https://buffer.com/).

9. ChatGPT Assistance:

> a. Automate content creation for blog posts and social media captions. b. Generate ideas for engaging and creative social media content. c. Draft email newsletters and promotional messages. d. Answer frequently asked questions from clients.

> With a limited budget and time, this marketing plan focuses on leveraging free tools and organic growth strategies. As your business grows, you can reinvest your revenue into expanding your marketing efforts and scaling your services.

In this instance, the plan is quite high level to give you the flexibility to decide how you wish to move forward with your own plan. But all you need to do with the output from Chat GPT is ask it to expand on the plan and help refine it for your specific needs.

> "I believe this artificial intelligence is going to be our partner. If we misuse it, it will be a risk. If we use it right, it can be our partner."

Masayoshi Son, Japanese Billionaire & Technology Entrepreneur

GROWTH

The following few chapters contain templates, top tips and copy and paste prompts for when your business is up and running.

CUSTOMER SUPPORT AND RETENTION

FOSTERING LOYALTY AND LONG-TERM RELATIONSHIPS

Outstanding customer support is essential for retaining customers and nurturing loyalty within your ChatGPT-based venture. In this section, we will offer a framework for devising an effective customer support strategy, guidance for pinpointing key customer touchpoints and areas that need enhancement, and tips on leveraging ChatGPT to elevate customer support and retention rates.

A robust customer support strategy serves as the foundation for building strong relationships with your customers and ensuring their satisfaction. By developing a strategy that addresses your customers' unique needs and preferences, you can create a positive, memorable experience that encourages repeat business and drives long-term growth.

Goals and Objectives

The first step in crafting your customer support strategy is to identify your goals and objectives. These may include improving response times, increasing customer satisfaction scores, or reducing churn rates. By setting clear, measurable targets, you can monitor the effectiveness of your customer support efforts and make data-driven adjustments to your approach.

Infrastructure

Next, evaluate your current customer support infrastructure and identify areas for improvement. This may involve assessing your support channels, staff training, and communication tools, as well as examining customer feedback and metrics to uncover pain points and opportunities for enhancement.

Pinpointing key customer touchpoints is crucial for understanding and optimizing the customer journey. To do this, create a worksheet that maps out each stage of the customer experience, from initial contact to post-purchase support. This exercise will help you identify areas where you can improve customer interactions and make a lasting positive impression.

Utilizing ChatGPT to bolster your customer support and retention efforts offers several advantages, including increased efficiency, personalized assistance, and the ability to scale your support operations. Some tips for incorporating ChatGPT into your customer support strategy include:

Chatbots and virtual assistants:

- Implement ChatGPT-powered chatbots or virtual assistants to provide real-time support and information, enabling customers to quickly find answers to their questions and resolve issues.

Personalized support:

- Use ChatGPT to tailor support responses based on customer data and preferences, creating a customized experience that demonstrates empathy and understanding.

Automated follow-ups and engagement:

- Leverage ChatGPT to send automated follow-up messages, check-ins, or satisfaction surveys, ensuring that your customers feel valued and supported even after their initial interactions.

Knowledge base and FAQ generation:

- Employ ChatGPT to create and maintain a comprehensive knowledge base or FAQ section, providing customers with easy access to information and reducing the burden on your support team.

In conclusion, exceptional customer support is critical for retaining customers and fostering loyalty in your ChatGPT-based business. By developing a well-structured customer support strategy, identifying key customer touchpoints, and harnessing the power of ChatGPT, you can elevate your customer support efforts and drive lasting success for your business.

TEMPLATE:

PLANNING AN EFFECTIVE CUSTOMER SUPPORT STRATEGY

Use this template to develop a customer support strategy that meets the needs of your customers and enhances their experience:

1. Set customer support goals:

- What do you aim to achieve with your customer support efforts (e.g., faster response times, increased customer satisfaction, reduced churn)?

- How will your customer support goals align with your overall business objectives?

2. Define your target audience:

- Who are your customers, and what are their needs, preferences, and expectations regarding customer support?

3. Determine support channels:

- Which support channels (e.g., email, live chat, phone, social media) will be most effective for addressing your customers' needs and preferences?

- How will you allocate your resources and budget across these channels?

4. Develop support resources:

- What resources (e.g., FAQs, knowledge base, tutorials) will you create to help customers find answers to common questions and issues?

5. Establish a support team structure:

- How will you structure your customer support team, including roles, responsibilities, and escalation paths?

- What training and resources will you provide to your support team members?

6. Measure and optimize performance:

- What key performance indicators (KPIs) will you track to evaluate the success of your customer support efforts?

- How will you use data and insights to inform future customer support decisions and improvements?

WORKSHEET:

IDENTIFYING KEY CUSTOMER TOUCHPOINTS AND AREAS FOR IMPROVEMENT

Complete the following worksheet to identify key customer touchpoints and areas for improvement in your customer support process:

1. Map your customer journey:

- Outline the key stages and touchpoints in your customer journey, from awareness and consideration to purchase and post-purchase support

2. Assess your current support performance:

- Evaluate your existing customer support efforts at each touchpoint, noting strengths and weaknesses

3. Identify areas for improvement:

- Highlight touchpoints where customers encounter issues or frustrations, and prioritize these areas for improvement

4. Develop improvement action plans:

- Create action plans for addressing the identified issues, including specific tasks, deadlines, and responsible team members

TOP TIPS:

USING CHATGPT TO BOOST CUSTOMER SUPPORT AND RETENTION RATES

1. Automate routine support tasks:

- Use ChatGPT to handle simple customer inquiries and tasks, freeing up your support team to focus on more complex issues

2. Create personalized support experiences:

- Leverage ChatGPT to tailor support interactions to individual customers, enhancing their experience and satisfaction

3. Draft response templates:

- Utilize ChatGPT to generate response templates for common customer inquiries, ensuring consistent and efficient support

4. Enhance self-service resources:

- Employ ChatGPT to develop and improve your knowledge base, FAQs, and other self-service resources, helping customers find answers quickly

5. Monitor and analyze customer feedback:

- Use ChatGPT to gather and analyze customer feedback, identifying trends and areas for improvement in your support process

By implementing a robust customer support strategy and effectively using ChatGPT to enhance support experiences, you'll improve customer retention rates and foster long-term loyalty in your ChatGPT-based business. In the next section, we'll discuss monitoring and analytics for your business.

CHATGPT SHORTCUT

Now for this section of the book it's important to note that the specific prompt you would need to use for ChatGPT is entirely dependent on the areas of improvement needed in your customer retention strategy. Obviously I can't cover every option here, so I have included an example as a template specifically to address a need to reduce churn and move existing customers away from a monthly subscription to an annual payment plan.

COPY AND PASTE PROMPT

Remember you will need to tailor this specific prompt to whatever issue needs resolving in your business once it's up and running. This is just an example.

> "Create a customer support strategy for my [business type] that offers [service/product] to [audience], I want to [problem to resolve] and [problem to resolve]."

PROMPT EXAMPLE

(USING THE TRAVEL MEDIA CO)

> "Could you create a customer support strategy for my online business that offers social media management to small travel companies please, I want to reduce churn and move customers away from a monthly model to an annual model."

CHATGPT'S RESPONSE

Certainly! To create a customer support strategy that will help reduce churn and encourage customers to switch from a monthly to an annual subscription model, consider the following steps:

1. Set clear expectations: Clearly communicate your service offerings and benefits to your customers, so they know exactly what they're getting with your social media management services. Highlight the benefits of moving to an annual subscription model, such as cost savings and priority support.

2. Offer personalized onboarding: When a new customer signs up, provide a personalized onboarding experience that caters to their unique needs. Assign a dedicated account manager to each client who will guide them

through the setup process, understand their goals, and develop a custom social media strategy for their travel company.

3. Implement a knowledge base and self-help resources: Create a comprehensive knowledge base with articles, FAQs, and video tutorials to address common questions and issues. This will help customers find solutions quickly and reduce the burden on your support team.

4. Provide timely and efficient support: Ensure your customer support team is well-trained and responsive, handling inquiries and resolving issues quickly. Offer support through multiple channels such as email, chat, and phone to cater to different customer preferences.

5. Regularly review and optimize performance: Monitor your clients' social media performance and provide them with regular re-

ports, highlighting successes and areas for improvement. Offer insights and suggestions to help them achieve their goals and demonstrate your expertise and value.

6. Engage with customers proactively: Regularly check in with your clients to gather feedback, address any concerns, and keep them informed about new features and updates. This will help you build a strong relationship and showcase your commitment to their success.

7. Offer incentives for annual subscriptions: Encourage customers to switch to an annual subscription model by offering incentives such as discounted pricing, exclusive features, or priority support. Clearly communicate the benefits and savings of the annual plan compared to the monthly plan.

8. Implement a referral program: Create a referral program that rewards existing cus-

tomers for referring new clients to your business. This will encourage customer loyalty and help grow your customer base.

9. Monitor and address churn: Regularly analyze churn rates to identify trends and potential issues. Reach out to customers who have canceled or downgraded their subscriptions to understand their reasons and identify opportunities to improve your services and win them back.

10. Continuously improve your services: Stay up-to-date with the latest social media trends and tools to continuously enhance your service offerings. By providing exceptional value to your customers, you'll increase their satisfaction and loyalty, ultimately reducing churn and promoting long-term commitments to your services.

By implementing these strategies, you

can improve customer satisfaction, reduce churn, and encourage more clients to opt for an annual subscription model.

"Intelligence is the ability to adapt to change."

Stephen Hawking, Theoretical Physicist

MONITORING AND ANALYTICS

HARNESSING DATA FOR CONTINUOUS IMPROVEMENT

To guarantee the success of your ChatGPT-based venture, it's crucial to monitor performance and make data-driven decisions. In this section, we will introduce a template for establishing a performance dashboard customized to your business, provide guidance for selecting and tracking essential KPIs, and share tips on leveraging ChatGPT-generated data to make informed business decisions.

A performance dashboard serves as a central hub for tracking and visualizing the key metrics that reflect the health and progress of your ChatGPT-based business. By designing a dashboard that is tailored to your unique goals and objectives, you can maintain a clear view of your business's performance and make timely, data-driven adjustments to your strategy.

The first step in setting up your performance dashboard is to identify the key performance indicators (KPIs) that align with your business objectives. These may include metrics related to customer acquisition, engagement, revenue, churn, and customer satisfaction. By selecting KPIs that are relevant and actionable, you can focus your attention on the most critical aspects of your business's performance.

Next, create a checklist for tracking and analyzing each KPI. This should include the data sources, reporting frequency, and benchmarks for each metric, as well as any associated targets or goals. Establishing a structured process for monitoring and analyzing your KPIs will help you stay informed about your business's performance and make data-driven decisions.

Leveraging ChatGPT-generated data can provide valuable insights into your business's operations and customer interactions, informing your decision-making process and driving improvements.

Some tips for harnessing ChatGPT-generated data include:

Sentiment analysis:

- Utilize ChatGPT to analyze customer feedback, reviews, and comments, gauging sentiment and uncovering trends or pain points that can inform your customer support and product development efforts.

Content performance:

- Monitor the engagement and conversion

metrics for ChatGPT-generated content, such as click-through rates, time on page, and conversion rates. Use this data to refine your content strategy and optimize your messaging for maximum impact.

Customer support metrics:

- Track ChatGPT-powered customer support interactions, such as response times, resolution rates, and satisfaction scores. Analyze this data to identify areas for improvement and optimize your support processes.

A/B testing:

- Employ ChatGPT-generated content and messaging in A/B tests to determine the most effective strategies and tactics for your marketing, sales, and customer support efforts.

In conclusion, monitoring and analytics play a vital

role in ensuring the success of your ChatGPT-based business. By setting up a performance dashboard tailored to your business, selecting and tracking essential KPIs, and harnessing ChatGPT-generated data, you can make informed, data-driven decisions that drive growth and lasting success.

TEMPLATE:

SETTING UP A PERFORMANCE DASHBOARD FOR YOUR BUSINESS

Use this template to create a comprehensive performance dashboard that monitors the key metrics of your ChatGPT-based business:

1. Identify key business objectives:

- Outline your primary business objectives (e.g., revenue growth, customer acquisition, customer satisfaction) to ensure your dashboard is aligned with your goals

2. Select relevant KPIs:

- Choose the most relevant KPIs to track for each business objective, ensuring a comprehensive overview of your performance

3. Determine data sources:

- Identify the data sources (e.g., ChatGPT API, Google Analytics, social media platforms) that will provide the necessary information for each KPI

4. Design your dashboard layout:

- Organize your dashboard in a logical and visually appealing layout, grouping related KPIs together and prioritizing key metrics

5. Implement data visualization:

- Use charts, graphs, and other visualizations to present your KPI data in an easy-to-understand and actionable format

6. Schedule regular reviews:

- Establish a routine for reviewing your dashboard, such as weekly or monthly, to monitor performance and make data-driven decisions

CHECKLIST:

SELECTING AND TRACKING ESSENTIAL KPIs

Consider the following KPIs when creating your performance dashboard, and select those most relevant to your ChatGPT-based business:

1.Revenue and profitability:

- Total revenue, net profit, average revenue per user (ARPU), and customer lifetime value (CLV)

2. Customer acquisition and retention:

- New customers acquired, customer churn

rate, and retention rate

3. Engagement and satisfaction:

- Average session duration, bounce rate, pages per session, and customer satisfaction score (CSAT)

4. Marketing and sales performance:

- Conversion rate, cost per acquisition (CPA), return on ad spend (ROAS), and email open and click-through rates

5. ChatGPT-specific metrics:

- API usage, token consumption, response time, and error rate

TOP TIPS:

LEVERAGING CHATGPT-GENERATED DATE TO MAKE INFORMED BUSINESS DECISIONS

1. Identify trends and patterns:

- Analyze ChatGPT-generated data to identify trends, patterns, and correlations that can inform your business strategy and decision-making

2. Evaluate ChatGPT performance:

- Monitor ChatGPT-specific KPIs, such as response time and error rates, to assess the performance of your integration and identify areas for improvement

3. Optimize content and marketing efforts:

- Use ChatGPT-generated insights to identify high-performing content and marketing tactics, and adjust your strategy accordingly

4. Enhance customer support:

- Analyze ChatGPT-generated customer feedback and support interactions to identify areas for improvement and implement necessary changes

5. Inform product development and innovation:

- Leverage ChatGPT-generated data to identify customer needs, preferences, and pain points that can guide your product development and innovation efforts

By setting up a comprehensive performance dashboard and effectively leveraging ChatGPT-generated data, you'll be well-equipped to make informed decisions and drive ongoing success for your ChatGPT-based business. In the next section, we'll discuss strategies for scaling your ChatGPT business.

CHATGPT SHORTCUT

Again as the dashboard and KPI's are going to be unique to your business. Below is an example of what you could use in ChatGPT to answer this question.

COPY AND PASTE PROMPT

Create a KPI dashboard for my [business type], which offers [product/service] to [target market]. The dashboard should include the following KPIs:

1. [KPI 1]: Track [metric/indicator] to measure [business goal/objective].

2. [KPI 2]: Monitor [metric/indicator] to gauge

[business goal/objective].

3. [KPI 3]: Analyze [metric/indicator] to uncover [trend/insight].

The data for these KPIs will be collected from [data source(s)]. I would like the dashboard to be presented in [table/visual/both] format.

PROMPT EXAMPLE

(USING THE TRAVEL MEDIA CO)

Create a KPI dashboard for my social media management business, which offers social media management services to small travel agencies. The dashboard should include the following KPI's:

1. Engagement rate: Track the number of likes,

comments, and shares on social media posts to gauge the level of engagement from the target audience.

2. Conversion rate: Analyze the percentage of website visitors who take a desired action, such as making a purchase or filling out a form, to determine the effectiveness of the social media strategy in generating leads and sales.

3. Response time: Track the time it takes for the business to respond to customer inquiries and comments on social media platforms to gauge the effectiveness of the customer support efforts.

The data from these KPIs will be collected from Google Analytics & Meta Analytics. I would like the dashboard to be presented in a table format.

TOP TIP: I found that GPT4 often refuses to do this, claiming that as an AI Language Learning Model, it can't create tables. In this instance using GPT3.5 seems to solve the problem instantly!

CHATGPT'S RESPONSE

As the response was in a table I've added this one as an image:

Again this is just an example and you'll need to add your specific KPIs into your prompt. And don't forget you can always ask for ChatGPT's recommendation on what KPIs it would recommend for your type of business.

REMINDER: As a thank you for buying this book. I have created a simple Copy & Paste PDF for you. You can get your copy from the QR code at the back of this book.

"I often tell my students not to be misled by the name 'artificial intelligence' - there is nothing artificial about it. AI is made by humans, intended to behave by humans, and, ultimately, to impact humans' lives and human society."

Fei-Fei Li, Stanford Univeristy Data Scientist

SCALING YOUR CHATGPT BUSINESS

EXPANDING HORIZONS AND ACHIEVING SUCCESS

The ability to scale effectively is a critical factor in the long-term success of any business, including ChatGPT-based ventures. As your business grows, exploring and capitalizing on opportunities for scaling becomes increasingly important to maximize your success, increase your market share, and maintain profitability. In this chapter, we'll delve into the importance of scaling and how to prepare your ChatGPT-based business for sustainable growth.

Scaling a business is about more than simply expanding your operations; it involves strategically aligning resources, processes, and strategies to support increased demand and reach new customers while preserving or enhancing efficiency and profitability. For ChatGPT-based businesses, this often involves not only growing your customer base but also finding ways to leverage the power of AI and machine learning to optimize your operations and develop innovative, scalable solutions.

Before embarking on a scaling journey, it's essential to establish a strong foundation. This includes having a clear understanding of your target market, a well-defined value proposition, and a robust business model that can support growth. Additionally, a strong team, efficient processes, and the right technology infrastructure are critical to successfully scaling your ChatGPT-based business.

One of the unique aspects of scaling a ChatGPT-based business is the potential to harness the power of artificial intelligence and machine learning to streamline operations, automate tasks, and enhance your product or service offerings. As you grow, consider how ChatGPT can be further integrated into your business processes, enabling you to scale more effectively and efficiently.

As you prepare to scale your ChatGPT business, it's important to recognize that scaling is an ongoing process that requires constant monitoring, adaptation, and optimization. Regularly evaluating your performance, identifying areas for improvement, and making data-driven decisions will help ensure that your business continues to meet the needs of your expanding customer base and remains competitive in an ever-changing market.

In the following sections of this chapter, we'll provide valuable guidance and resources to help you assess your business's scalability potential, identify areas for growth and expansion, and share tips on scaling a ChatGPT business effectively and efficiently. By carefully planning and executing your scaling strategy, you can propel your ChatGPT-based business to new heights and drive lasting success.

WORKSHEET:

ASSESSING YOUR BUSINESS'S SCALABILITY POTENTIAL

Complete the following worksheet to evaluate your ChatGPT-based business's scalability potential:

1. Market size and growth:

- Assess the size of your target market and its potential for growth to determine the expansion opportunities for your business

2. Product or service scalability:

- Evaluate whether your ChatGPT-based offering can be easily scaled to meet the increasing demands of a larger customer base

3. Operational efficiency:

- Analyze your current processes and workflows to identify potential bottlenecks or inefficiencies that could hinder your ability to scale

4. Financial capacity:

- Assess your business's financial resources and ability to support the costs associated with scaling

5. Team and talent:

- Evaluate your team's capacity and capabilities to handle the increased workload and responsibilities that come with scaling your business.

CHECKLIST:

PINPOINTING AREAS FOR GROWTH AND EXPANSION

Use this checklist to identify areas for growth and expansion in your ChatGPT-based business:

1. New markets or customer segments:

- Explore opportunities to expand your offering to new markets or customer segments, either geographically or demographically

2. Product or service extensions:

- Consider enhancing your ChatGPT-based offering with additional features or complementary services

3. Partnerships and collaborations:

- Identify potential partnerships or collaborations that could help you reach new customers, access new markets, or enhance your offering

4. Marketing and sales expansion:

- Look for opportunities to scale your marketing and sales efforts, such as increasing your advertising budget, adopting new marketing channels, or expanding your sales team

5. Process automation and optimization:

- Evaluate opportunities to streamline and automate your business processes to improve efficiency and support growth

TOP TIPS:

SCALING YOUR CHATGPT BUSINESS EFFECTIVELY AND EFFICIENTLY

1. Prioritize your scaling efforts:

- Focus on the most impactful growth opportunities first, based on your assessment of your business's scalability potential

2. Monitor performance closely:

- As you scale your business, keep a close eye on your KPIs and other performance metrics to ensure you're maintaining quality and achieving your objectives

3. Optimize ChatGPT performance:

- Continuously refine and optimize your ChatGPT integration, focusing on improving performance, reducing costs, and enhancing the customer experience

4. Invest in team development:

- As your business grows, invest in training and professional development for your team members to ensure they're equipped to handle the increased workload and responsibilities

5. Maintain a strong company culture:

- As you scale, focus on preserving and nurturing your company culture to ensure continued alignment with your core values and mission

By assessing your business's scalability potential and strategically pursuing growth opportunities, you can effectively and efficiently scale your ChatGPT-based business to achieve long-term success.

CHATGPT SHORTCUT

Again the exact prompt you'll need will depend on where your business is and what areas you would like to scale up in. I have included a general example of a prompt to scale post year 1, as an example:

COPY AND PASTE PROMPT

Please write me a plan to scale my [business type] for [audience] at [stage of business]

EXAMPLE PROMPT

"Please write me a plan to scale my social media management business for small travel agencies post year 1."

CHATGPT'S RESPONSE

Sure! Here's a plan to scale your social media management business for small travel agencies post year 1:

1. Expand your services: After one year of working with small travel agencies, you likely have a good idea of what services are in high demand. Use this knowledge to expand your services beyond social media management. For example, you could offer website design, SEO services, email marketing, or content creation.

2. Develop a referral program: Referrals are a great way to bring in new business. Create a referral program that rewards clients for referring new customers to your business. This could be a discount on your services, a free consultation, or another incentive that makes sense for your business.

3. Build partnerships: Partner with other businesses that serve your target audience. For example, you could partner with a travel agency or a hotel to offer joint promotions or packages.

4. Invest in advertising: Once you have a solid client base and a good understanding of what services are in demand, consider investing in advertising to reach a wider audience. This could include social media advertising, Google AdWords, or other forms of online advertising.

5. Hire additional staff: As your business grows,

you may need to hire additional staff to manage the workload. Consider hiring a part-time or full-time social media manager, graphic designer, or content creator to help you meet the demand for your services.

6. Leverage automation: As your business grows, you'll want to find ways to work more efficiently. Leverage automation tools like Hootsuite or Buffer to streamline your social media management tasks.

7. Develop a training program: If you plan to hire additional staff, develop a training program to ensure that everyone on your team is on the same page. This will help ensure that your clients receive consistent, high-quality service.

8. Attend industry events: Attend industry events and conferences to stay up-to-date on the latest trends and best practices in social media management and marketing. This

will help you stay ahead of the curve and offer the best possible service to your clients.

By following these steps, you can scale your social media management business for small travel agencies post year 1. Remember to stay focused on your clients' needs and offer exceptional service to keep them coming back for more!

REMINDER: As a thank you for buying this book. I have created a simple Copy & Paste PDF for you. You can get your copy from the QR code at the back of this book.

CONCLUSION

So there we have it! We are at the finale of *The ChatGPT Millionaire Blueprint*, let's take a moment to revisit the vital components that will propel your ChatGPT-based business toward unparalleled heights and pave your way to becoming a millionaire:

1. **Market Research and Niche Selection:**
   Unlock the secrets to your target audience, niche, and the market landscape, laying the unshakable foundation for your thriving venture.

2. **Business Model and Revenue Streams:**
   Craft a robust business model and discover lucrative revenue streams to catapult your

business into profitability.

3. **Building Your Platform:**

Design an enthralling website or app that embodies your brand essence and captivates your target audience, marking your territory in the digital world.

4. **ChatGPT Integration and Optimization:**

Strategically integrate and optimize ChatGPT to unleash the full potential of this transformative AI powerhouse.

5. **Content Creation and Automation:**

Devise a magnetic content strategy and harness ChatGPT to generate compelling content that enchants your audience and drives tangible results.

6. **Marketing and Promotion:**

Engineer a cutting-edge marketing plan and leverage ChatGPT to amplify your reach, skyrocketing your business growth.

7. **Customer Support and Retention:**
   Create a customer support masterpiece and employ ChatGPT to elevate customer interactions, fostering loyalty and retention.

8. **Monitoring and Analytics:**
   Establish an insightful performance dashboard and tap into the wisdom of ChatGPT-generated data to steer your business with confidence.

9. **Scaling Your ChatGPT Business:**
   Evaluate your business's boundless scalability and seize growth opportunities that catapult you toward enduring success.

Now, with the knowledge and tools unveiled in this book, you're ready to forge your own ChatGPT-based business empire.

To take your journey to new heights, and if you are looking for unparalleled support, expert mentorship and an arsenal of insights then get in touch with me at my website www.businessmaven.io -

where I offer a host of services to help you get the most our of your business venture.

The stage is set, and the world eagerly awaits your triumph. Fortune favors the bold, and your moment has arrived!

BONUS - PROMPTS

As a thank you for buying this book, you'll also be getting a bonus of all the prompts used in an easy to use copy and paste pdf. Simply scan the QR code below to get access. Or you can visit https://www.businessmaven.io/chatgpt

If you have any questions or queries, or if you need

further support please make sure to use the contact page on my website: https://www.business-maven.io/contact

Hey there,

Thank you so much for finishing this book!

Did you know that positive reviews from awesome customers like you help this book to be shown to other people just like you and help them to feel confident about also reading this book.

Could you take 60 seconds to go to Amazon and share your experiences?

I will be forever grateful.

Thank you in advance for helping us out!

Business Maven

ABOUT AUTHOR

Business Maven is a veteran Fortune 500 Consultant, Business Coach & AI Strategist.

Over the years she has worked with companies like Samsung, o2, Virgin, Google, Apple & Sony. She now travels the world helping startups, entrepreneurs and corporates alike excel in the AI age.

Her work also covers digital transformation, AI integration as well as strategies for success and mindset resilience for entrepreneurs. Some of her mindset work has been downloaded over 55,000 times and she has a growing social media presence since turning her attention to writing books.

For more on her work, visit www.businessmaven.io

With Thanks

I simply wish to thank my mum and dad, without your love, support and belief in me none of this would be possible.

I am eternally grateful to you both

xxx